Sustainable Development in Rural China

Bingtao Qin

Sustainable Development in Rural China

Field Survey and Sino-Japan Comparative Analysis

Springer

Bingtao Qin
Business School
University of Shanghai for
Science & Technology
Shanghai, China

Sponsored by Chinese Fund for the Humanities and Social Sciences (本书获中国哲学社会科学基金资助)

ISBN 978-3-662-46475-5 ISBN 978-3-662-46476-2 (eBook)
DOI 10.1007/978-3-662-46476-2

Library of Congress Control Number: 2015936194

Springer Heidelberg New York Dordrecht London

Printed on acid-free paper

Springer-Verlag GmbH Berlin Heidelberg is part of Springer Science+Business Media (www.springer.com)

Abstract

With industrialization and the rapid economic growth, environmental pollution, ecological destruction, and other issues have become increasingly prominent in China. Due to its large population size, vast land area, and the impact that its development has on the global economy, researching sustainable development in China is even an important issue for the whole world. However, the focus of China's development should be in rural areas.

First of all, China's rural population accounts for most of the total population and most of the poor are living in rural China. To meet the demand for food in such a huge population, modern agricultural production has made heavy use of pesticides and fertilizers and this has caused significant damage to the environment in rural areas. Secondly, due to the industrialization of rural China, industrial pollution damages the environment in rural areas together with modern agricultural production. In addition, keeping the surplus labor migrate from rural regions is an important issue not only for the economic development but also for reducing the pressure on the rural ecological system. Finally, due to the lack of a social system, the aging problem caused by labor's flowing-out has put great pressure on rural China residents.

These issues are important factors in impeding sustainable development in rural China. In this paper, they are divided into five topics: ecological agricultural production, industrial pollution, renewable energy, labor mobility, and the provision of a welfare system, which have been outlined in five chapters. The author believes that the path to sustainable development for rural China should be in an integrated and systemic way in three views of the economy, ecology, and the social system. As to the analysis methods, this paper adopts a method which combines both field survey and a comparative analytical approach. These should be called the originalities of this book. With the analysis, this book has the following main conclusions:

1. It is necessary to improve modern agricultural production methods, and the choice to employ more ecologically friendly forms of agriculture is a way forward in ensuring a brighter future.

2. To protect the environment of rural regions, the Chinese government should control industrial pollution with environmental legislation, law enforcement, and corrections to the bureaucratic evaluation system.
3. Ensuring a sustainable path to energy security, China should break its reliance on fossil-based energy structure but develop and utilize renewable energy sources such as wind power.
4. So as to ensure the surplus labor's migration, reforming the household registration system would play a positive role, as well as other policies.
5. To resolve the aging and health problems in rural areas, China should establish a compound old age security system and urban-rural unified health insurance system.

Given on the analysis previously, this book has identified some paradigmatic scenarios for the development of rural China: a default scenario, a sustainable scenario, and some other scenarios depending on how the government pays attention to these fields mentioned previously. The earlier the government adopts reasonable policies, the easier to achieve the sustainable situation in rural China. In order to fully achieve the end result described for sustainable scenario, efforts should begin immediately in a wide scale in rural areas. Research on sustainable development in rural China should be carried out over a long period by many more researchers. This book is just one attempt at addressing the issues raised and only a beginning in my own study into such issues which I hope to pursue in the foreseeable future.

Acknowledgments

This book is based on my thesis for Ph.D. of economics in Niigata University. I wish to express my sincere gratitude and appreciation to my main supervisor Professor Fumiaki Toudou. I wish to thank him for his expert guidance and advice throughout my entire course of study. I wish to extend my appreciation to Professors Keniti Haga and Takeshi Yamazaki for all their valuable advice and help. I acknowledge all the assistance that I received from Professors Yoshin Sugahara, Takashi Fujii, Yuuki Mizoguti, Koji Takamiya, Yoji Koyama, Takeshi Kuniya, and the staffs of the Graduate School of Modern Society and Culture, who have always offered me great assistance during my studies at Nigata University. In addition, I thank associate professors Dene Croxford and George O'Neal who checked over my draft for me with regards editing language.

I am indebted to Professor Chen Shiyi who recommended my thesis to Springer and Toby Chai who put a lot in the publishment. Special thanks go to Chinese Fund for the Humanities and Social Sciences (CFHSS), this book would not have been possible without its support. I wish to thank everyone who was involved in interviews and answering questionnaires as it was with their cooperation that I was provided with the first-hand data which I required for this book. I also thank my friends for giving me their encouragement and my host families for their kindness and hospitality toward me during my homestays in Yoshida. Finally, I wish to express my gratitude toward my parents and to the rest of my family, including my wife and children, for their unconditional love and support.

Contents

Chapter 1
Introduction

1.1 The Background of Research

Since the industrial revolution, human society has been changing from a traditionally agricultural based civilization to a modern industrial one.[1] Some European countries and America have almost completed this process of industrialization and have become fully developed industrialized societies. During the period of development, these economies have achieved rapid economic growth and their living standards have been increased significantly. Since then, and especially since World War II, developing countries have viewed developed countries as a template by which to implement their industrial development policies.[2] Therefore, industrialization and economic growth have become the recurrent theme for development throughout the world. However, at the same time, an excessive expansion of the population, environmental pollution, ecological destruction and other issues have become increasingly prominent throughout the world. These issues are threatening the future of our human society. We should re-examine the current economic and social development model and explore a more sustainable path to development.

The traditional concept of development only included economic growth and viewed the gross domestic product (GDP) as the means by which to measure it. Based on this understanding, persuading the growth of GDP became the sole objective of governments' economic policy. However, this economic accounting system has many defects such as the exclusion of unmarketable activities which lie outside official statistics, etc. Especially, it does not take the environmental factors into account. This accounting system ignores not only the scarcity of resources, but also the destruction of the environment quality caused by pollution.[3]

---

[1] Deng and Zhang (2005).

[2] Yang (1999).

[3] Xie (2011).

B. Qin, *Sustainable Development in Rural China*,
DOI 10.1007/978-3-662-46476-2_1

Rapid economic growth is often co-existent with huge energy consumption and the cause of serious environmental damage. We have seen that energy consumption in the world is rising sharply and the world energy structure is relying mainly on fossil energy, coal, oil and natural gas which will be exhausted in the near future. It is estimated that oil reserves can only be maintained until 2031, natural gas until 2050, and coal 2100.[4] At the same time, it is needless to say that human activities, especially industrial production are the most serious cause of environmental pollution. It has been estimated that over one million tons of toxic substances are emitted daily in the earth's atmosphere.[5] These harmful substances are contaminating ecosystems such as soil, water resources, and damage to human health directly.

After extensive debate and researches, the international community has been realizing that only pursuit of material growth is limited. Economic growth only is limited to reflect the development and progress of human society. Economic growth refers to the increase in the amount of material, while development is qualitative improvement.[6] Economic growth is an essential means of development, but not the goal.[7] Economic development should be in harmony with the environment, and not at its expense.

In fact, the issue of relationship between human society and the natural environment has always been associated with the history of civilization. Economic development and environmental issues have also emerged in the development of agriculture. The earliest record in China can be traced back 4,000 years ago. Based on the ancient texts, the King of DaYu[8] educated his people in the art of cultivating young trees avoiding their annual deforestation in the month of March, and protecting and preserving young fish by avoiding fishing in the month of June. On population issues, during the Warring States period, Han Fei[9] developed the theory that contrary to absolute wealth, a relative amount of wealth could be achieved in comparison to the size of the population. This is reasonable because the population can grow exponentially.

The impact of human society on the environmental depends not only on population size but also on the technical advancement. In a traditional agricultural society, because the population size was relatively small, the impact of agricultural production on the environment was weak and the environmental damage caused by human economic activities was regional and on a small-scale. However, with the growth in population and industrialization, the environmental impact of human activities has become more significant and the effects are longer lasting. Further-

[4] Meadows (1992).

[5] Solid Wastes Management Center of Henan Province (2011).

[6] Daly (1991).

[7] World Bank (1992).

[8] The king of China about 4,000 years ago, he is famous for management of the Yellow River flood.

[9] A very famous philosopher in China's Warring States Period.

more, with an ever more internationally based division of production and economic globalization, environmental issues have become a worldwide concern. There are an increasing number of experts who think that there should be an adjustment in the relationship between economic activity and its environmental impact upon an agricultural society which cannot be suitable in order to develop a modern industrial society. Searching for a path of sustainable development path has become the only option for us now.

1.2 Overview of Prior Studies

"Silent Spring"[10] is recognized as the first work which advocated the sustainable path to development.[11] This work pointed out that the use of pesticides irreparably causes harm to the human body and can even the entire ecosystem. In 1972, the Academic Community of Rome Club published its report entitled "Limits to Growth".[12] The publication pointed out that the globe will have reached its limit to growth in the twenty-first century due to food shortage, depletion in energy resources and environmental destruction. These works have been recognized as awaking the world to the risks posed by conventional thinking in terms of economic growth and have promoted a deeper understanding of sustainable development. The idea of sustainable growth in agriculture has gradually been taken more seriously.

In 1980, the International Union for Conservation of Nature (IUCN)[13] developed the "World Conservation Strategy"[14] which proposed the concept of sustainable development for the first time. In 1987, the World Commission on Environment and Development presented a report entitled "Our Common Future".[15] This report noted that because we have mostly concerned about economic development in the past, we are feeling greater urgency about the pressure place upon the ecosystem. At the same time, this document defined sustainable development as improving the quality of human life, without exceeding the carrying capacity of the ecosystem.[16] In 1992, the United Nations Conference on Environment and Development[17] adopted

[10]Rachel Carson (1962).

[11]Ibid; see note 1.

[12]Rome Club (1972).

[13]IUCN was founded in 1948; its aim is seeking to work with its members to achieve development that is sustainable and providing a lasting improvement in the quality of life for people all over the world. http://www.iucn.org/, accessed on 2012.1.6.

[14]IUCN (1980).

[15]World Commission on Environment and Development (1987).

[16]Ibid; see note 15.

[17]United Nations, Conference on Environment & Development was held in Rio de Janerio, Brazil, from 3 to 14 June, 1992.

the "Agenda 21[18]" proposing the establishment of new ideas and new concepts for sustainable development. This meeting was considered a milestone in the history of human development. From then on, sustainable development has become the generally accepted worldwide consensus. It illustrates a great leap in thinking to change from the traditional concept of development to one which is sustainable.

In China, after the United Nations Conference on Environment and Development, the government has formulated its "White Paper on Population, Resources, Environment and Development in the twenty-first Century"[19] in 1994. In this document, the government views sustainable development as the goal of China's economic development and has incorporated it into the country's long-term planning for the future. Subsequently in 1997, the CPC 15th meeting[20] determined the sustainable development strategy for China's modernization strategy.

However, the focus of China's development should be in rural areas. This is because, first of all, China's rural population accounts for 64 % of the total population[21] and most of the poor are living in rural China, so eradicating rural poverty is one of the basic goals of development in China. Secondly, in order to develop the economy and improve people's living standards, China has always industrialization as crucial ever since the establishment of the country. Thirdly, because of the huge surplus of labor force in rural China, this has provided a sufficient labor supply necessary for industrialization. On the contrary, the outflow of the labor force from rural areas has caused ageing problem in those populations which remain. Due to the lack of a social system, an ageing problem has put great pressure on rural China's rural social security system. Then again, to meet the demand for food in such a huge population, modern agricultural production has made heavy use of pesticides and fertilizers. Together with industrial production, this has caused significant damage to the environment in rural areas. These issues are important factors in impeding sustainable development in rural areas. We can say that if these issues cannot be resolved, neither can China's real sustainable development be maintained. Therefore, I have carried out research into the sustainable development of rural China as part of this book.

There are many prior studies on the development of rural China. However, the studies of sustainable development in rural areas have often only focused on a single field. Concerning the surplus of labor and the problems in transference from the countryside to the cities, Cai,[22] Wang,[23] and Nie[24] have made numerous studies.

[18]United Nations, Department of Economic and Social Affairs, Division for Sustainable Development (1992).

[19]China State Council (1994).

[20]1997.9.12–18, held in Beijing.

[21]National Bureau of Statistics of China (2000).

[22]Cai (2007).

[23]Wang (2008).

[24]Nie (2006).

Zhang[25] and Su[26] in particular, have researched the issue of an ageing population the rural pension system in China. Due to the ageing issues, pension insurance and health insurance are the most import issues in rural China. Liu,[27] Zheng,[28] and Jin[29] are all the experts in this field. As for eco-friendly agriculture, Li[30] has done research into this over many years and can be viewed as its expert. Chen[31] and Tao[32] have studied results in rural energy consumption including actual energy and green energy technologies. To research the sustainable development in rural China, these aspects are in closely related to each other. In other words, any study should be comprehensive and systematic, as just focusing on one or a few sides of the argument will not be effective in dealing with the overall problem.

There are far too few comprehensive studies on sustainable development in rural China. The research group at the Chinese Academy of Sciences has published a yearly report entitled "Report on Sustainable Development Strategy of China[33]" since 1999. These reports have mostly concentrated on topics of urbanization and the development of urban areas and rarely mention the rural areas. There are some scholars who are trying to research rural sustainability from the aspect of population, environment and resources, such as Fu,[34] and Zhong,[35] but the results are far too simplified and not very comprehensive. Even some scholars such as Zhou[36] and Lin[37] have produced more detailed studies and these studies have adopted a theoretical method which lacks any empirical research.

In order to make up the lack which exists in previous researches, this book will attempt to research sustainable development in rural China in an integrated and systemic way. In addition, I will attempt to sort out comprehensive factors associated with rural development and study the main factors towards sustainable development. Moreover, this book adopts a method which combines both theoretical and empirical research. Furthermore, in some topics which are commonly with the developed countries, I have taken a more comparative analytical approach. These should be called the originalities of this book.

---

[25]Zhang (2005).

[26]Su (2010).

[27]Liu (2004).

[28]Zheng (2011).

[29]Jin (2011).

[30]Li (2003).

[31]Chen (2008).

[32]Tao and Li (2007).

[33]Institute of Sustainable Development Strategy Research (1999–2010).

[34]Fu (2001).

[35]Zhong (2007).

[36]Zhou (2003).

[37]Lin (2005).

1.3 Explaining Sustainable Development

There are many concepts concerning sustainable development in the worldwide areas covering aspects at an international, regional, local and sector-specific level. One of them is from "World Natural Conservation Strategy"[38] put forward by the International Union for the Conservation of Nature[39] which declares that the basic relationship between all natural, social, ecological and economic factors as well as the process of using natural resources should be examined so that the sustainable development of the world can be assured.[40] The most widely mentioned concept is from "Our Common Future"[41] published by World Commission on Environment and Development in 1987. This report defined sustainable development as "development that meets the needs of the present generation and without hazarding the ability of future generations to meet their needs[42]". The definition focused on the intergenerational relationship of the whole debate on developmental sustainability and emphasized the fact that if problems cannot be resolved within one generation, then the whole model for sustainable development is not realistic.

In China, the government has formulated a document entitled "China's Agenda 21 – a White Paper on Population, Environment and Development in the twenty-first Century"[43] in 1994. This document summarized sustainable development as determined by five major components in terms of five components of population, society, economics, the rational use of resources and environmental protection. In September 1995, the Fifth Plenary Session of the 14th CPC Conference adopted "the Recommendations for Drawing up the Ninth Five-year Plan of National Economy and Social Development and the 2010 Vision".[44] This document clearly stated that to achieve sustainable development, it is necessary to control the population, to conserve resources and to protect the environment.

In the traditional study of modern economics, it is commonly used research methods to start from the constraint subject. In order to research the rural issues, the constraint subject is the reality of rural China and it is important to focus on it. Figure 1.1 shows the reality of rural regions.

First of all, there is a large agricultural population living in rural areas. Secondly, due to a limited land area per capita, there is a considerable surplus in labor, and agricultural production is mostly based on family-based operations. Also, as industrialization has spread from the city and coastal areas to rural and inland regions, energy security and industrial pollution problems have thus occurred.

[38]Ibid; see note 13.

[39]Ibid; see note 12.

[40]Pearce and Jeremy (1993).

[41]Ibid; see note 14.

[42]Ibid; see note 15.

[43]Ibid; see note 17.

[44]The Fifth Plenary Session of the 14th CPC Conference, 1995.9.28.

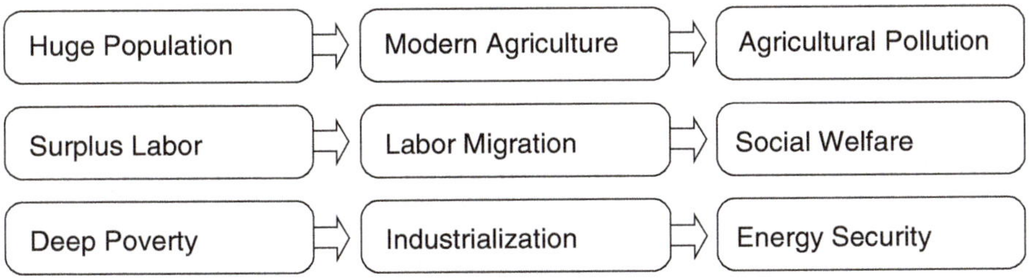

Fig. 1.1 The reality of rural China

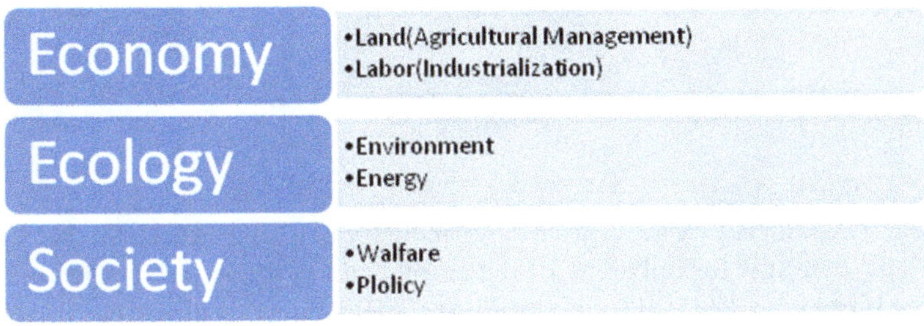

Fig. 1.2 Three aspects of sustainable development

Moreover, because of the outflow of labor force from rural areas, this has caused an ageing problem and a lack of social pension in rural areas. The ageing problem has in turn put great pressure on rural China's social security system.

Because of the differences between the urban and rural regions in terms of economic development, the ecological structure of the environment and the social security system, China's sustainable development should not only be intergenerational (vertical) but also be regional (horizontal). In addition, it is the integrated relationship between the economic, ecological and social fields that need to be looked at carefully and considered as showed in Fig. 1.2. Therefore, this book attempts to examine sustainable development rural China in three areas including economy, ecology and society. Furthermore, in regards to the previous three fields, I have selected five elements to be looked at. As Fig. 1.3 showed, these include agricultural production, labor migration, industrial pollution, and the social welfare system.

1.4 The Book Structure and Research Methods

In this book, the previously mentioned five topics will be divided into the five chapters. First of all, China has a population of over 1.3 billion and the demand for food has thus put huge pressure on agricultural production. To meet this demand for food, the heavy use of pesticides and fertilizers has improved land productivity

Fig. 1.3 Five main topics of sustainable development

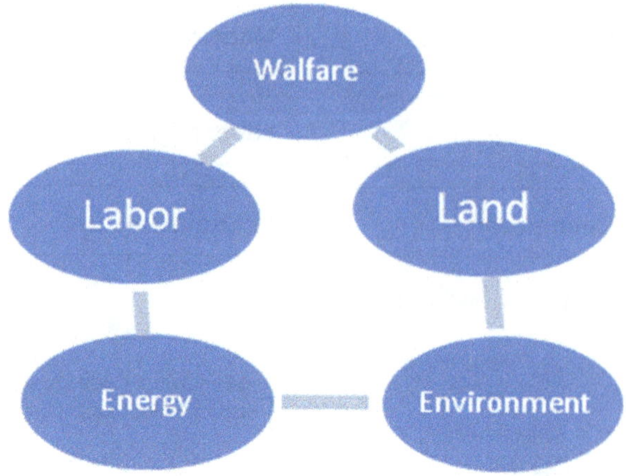

significantly. A Survey by the Agricultural Ministry of China has shown that China's arable land accounts for only 7 % of the arable land in the world, but fertilizer use accounts more than 40 % of total world use. The average fertilizer use in China is about 400 kg/ha and is far higher than the recognized safe quantity of 225 kg/ha in developed countries.[45] At the same time, the use of pesticides in China's agriculture has reached about 1.3 million tons annually. However, about 1/3 of the pesticides can be absorbed by crops and the left has been left to pollute 93 thousand square kilometers of arable land.[46] The monitoring of vegetable wholesale market over 16 main cities in 2002 showed that the overall detection rate of pesticide was 20–60 %.[47] It is far beyond the corresponding detection rate for developed countries. Fertilizers and pesticides not only directly threaten human health, but also pollute the soil and water sources while damaging eco-systems. Therefore, I will analyze the ecological agriculture in Chap. 2.

In order to address poverty, China has insisted on industrializing its economy ever since the foundation of the county 60 years ago. Especially after economic reform and the opening up policy of 30 years ago, China's industrial development has been developing rapidly. However, low-tech extensive industrialization has been very damaging to the environment. Especially in recent years, industrial pollution has also transferred to rural regions. As the pollution control system in rural China is weak, environmental pollution has become a very serious problem. There are more than 300 million rural residents who do not have clean drinking water and the mortality rate for cancer caused by environmental pollution in rural China has sharply increased.[48] If appropriate measures are not taken soon;

[45]Du (2009).

[46]Green Kunming (2010).

[47]Su (2005).

[48]Ibid; see note 47.

environmental pollution will seriously hamper the sustainable development of rural areas. Therefore, I have examined the question of the industrial pollution in Chap. 3.

As industrial development and the living standards of rural residents improve, electricity consumption has increased and this has put pressure on China's electricity supply capacity. Because the energy structure of China is so heavily reliant on fossil energy, such as coal and oil, this is not only the main cause of serious air pollution and acid rain, but also can be not sustainable due to the limit of supply of fossil energy sources. Therefore, the development and use of renewable energy is an important issue and are which will be further examined in Chap. 4.

In rural China, because of the limited amount of arable land, there has been a huge surplus in the labor force. The rural labor force accounted for 60 % of the whole national workforce in 2009, and the surplus in labor is estimated to be about 170 million in 2010.[49] The problem of rural labor migration is an important issue for sustainable development. The sheer number of experts and studies into labor migration problems are far too numerous to mention. Among these theories, Lewis's two-sector theory is recognized as suitable in evaluating China's actual situation and has been widely accepted by Chinese scholars such as Cai.[50] However, because the labor force shortage has recently emerged in coastal cities, many scholars believe that China's economy has reached the Lewis turning point. In Chap. 5, I will examine this issue.

Inter-regional labor migration has caused an ageing problem in rural areas. As shown in Chap. 6 Sect. 6.1, compared with 1982, the ageing rate in rural areas has increased from 5.0 % to 7.50 %. The ageing problem in the cities has also increased from 4.68 % to 6.67 % in the same period. This shows that the ageing problem in rural China is not only the most serious problem to affect the countryside but is also the fastest developing one. Because most parts of the rural regions have not established a social ageing insurance system, and the new rural cooperative medical care system is still in its pilot phase, compared to the cities, the ageing problem in rural areas has brought greater added pressures. In Chap. 6, I will examine the pension system in rural China.

This book will discuss the main problems affect China's developmental sustainability by comparing research and field survey methods. With regards eco-friendly agriculture and a social pension system, and given that Japan is both representative of a developed country and has features more closely related to China, the examination of Japanese research methods has been adopted. As for industrial pollution, I have selected a chemical plant and have carried out field research into the groundwater contamination problem. In order to look into renewable energy, I chose a wind power electricity generator to show that wind power is one of the sustainable energy sources that can be used in the future in China. Concerning the labor migration issue in Chap. 5, I have empirically surveyed 300 migrant workers with questionnaires in Shenzhen and analyze it in this book.

---

[49] Li (1997).

[50] Cai (2008).

References

Cai F (2007) Flowing population in transition China, 2nd edn. Social Sciences Academic Press, Beijing

Cai F (2008) Lewis turning point: a coming new stage of China's economic development. Social Science Press, Beijing

Carson R (1962) Silent spring. Houghton Mifflin Company, Boston

Chen YQ (2008) The use of new energy in rural China. Chinese Social Publisher, Beijing

China State Council (1994) China's agenda 21: white paper on population, resources, environment and development in the 21st century. China Environmental Press, Beijing

Daly H (1991) Steady-state economics, 2nd edn. Island Press, Washington, DC

Deng HB, Zhang Y (2005) Population resources and environmental economics. Science Press, Beijing, p 3

Du YY (2009) Thinking about China's food safety issues. Economic Information Daily, 25 Sept 2009

Fu C (2001) Research on the reform and major issues of rural China. Shanxi Economic Press, Taiyuan

Green Kunming (2010) Three types of environmental pollution in rural China. www.greenkm.org/html/200710/27/20071027154240.html. Accessed 9 July 2011

Institute of Sustainable Development Strategy Research (1999–2010) Report on sustainable development strategy of China. Science Press, Beijing

IUCN (1980). http://data.iucn.org/dbtw-wpd/edocs/WCS-004.pdf. Accessed 16 Aug 2011

Jin J (2011) Study of the medical insurance system in rural China. Zhejiang Business University Press, Hangzhou

Li HH (1997) The transfer problems of rural surplus labor on the next 15 years. Economic Theory and Economic Management Publishing House, Beijing, p 82

Li WH (2003) Ecological agriculture: the theory of sustainable agriculture and practice in China. Chemical Industry Press, Beijing

Lin FQ (2005) Research on the sustainable development of rural China. Chinese Market Press, Beijing

Liu HL (2004) New theory of population economics. China Economic Publishing House, Beijing

Meadows DH (1992) Beyond the limits. Universe Publishing Co., New York, p 163

National Bureau of Statistics of China, Fifth Population Census of China (2000) http://www.stats.gov.cn/tjgb/rkpcgb/. Accessed 4 Aug 2011

Nie HL (2006) Urban-rural relations in Western China. China Social Sciences Publishing House, Beijing

Pearce DW, Jeremy JW (1993) World without end: environment and sustainable development. Oxford University Press, New York, p 92

Rome Club (1972) The limits to growth: a report for the club of Rome's project on the predicament of mankind. Universe Publishing Co., New York

Solid Wastes Management Center of Henan Province. http://www.hnep.gov.cn/gfgl/tabid/683/InfoID/20376/Default.aspx. Accessed 16 Dec 2011

Su Y (2005) The environmental pollution during the process of modernization in rural areas. China Economic Times, 29 Nov 2005

Su CH (2010) Economic impact of aging and selection of pension. Economic Science Press, Beijing

Tao ZP, Li MY (2007) What is the limit of Chinese coal supplies: a TELLA model of Hubbert peak. Energy Policy 35(6):3145–3154, http://www.sciencedirect.com/science/article/pii/S0301421506004290. Accessed 13 June 2011

United Nations (1992) Department of Economic and Social Affairs, Division for Sustainable Development, 1992, Agenda 21. http://www.un.org/esa/dsd/agenda21/res_agenda21_00.shtml. Accessed 7 Jan 2012

Wang P (2008) Research on the transfer of surplus labor from rural China. Northeastern Financial University Press, Dalian

World Bank (1992) World development report 1992 – Development and the environment. Oxford University Press, Oxford/New York

World Commission on Environment and Development (1987) Our common future. Oxford University Press, Oxford

Xie HY (2011) Research on the product pricing mechanism with the environmental costs of resource, Institute of Economics System and Management National Development and Reform Commission. http://www.china-reform.org/?content_188.html. Accessed 9 Oct 2011

Yang YY (1999) Population resources and environmental economics. China Economic Publishing House, Beijing, p 11

Zhang YG (2005) China's pension system under the background of population ageing. Southwest University of Finance and Economics Press, Chengdu

Zheng GC (2011) China's social security reform and development strategies (old-age insurance volume). People's Publishing House, Beijing

Zhong SY (2007) Population resources and environmental economics. China Social Publishing House, Beijing

Zhou WB (2003) Sustainable development of rural China during the process of modernization. Hunan People's Publishing House, Changsha

Chapter 2
Sustainable Agricultural Development in Rural China: The Way of Addressing Agricultural Pollution

2.1 The Production and Pollution of Agriculture in Rural China

In the last century, the world's population grew rapidly. Especially after World War II, the population increased from about 2.5 billion then to 6.8 billion[1] which it is now. The rapid population growth caused a rapid demand in the growth of agricultural products. To meet this continuously increasing demand, agricultural production was under tremendous pressure. Given this background, so-called modern agriculture made wide use of chemical fertilizers and pesticides and mechanization in agriculture was promoted as the means by which agricultural production would be improved significantly.

In most industrially developed countries, the shortage of agricultural products has been resolved by modern agricultural production methods. Furthermore, based on the vast land area, former colonial countries, such as United States and Australia have become the world's major exporters of agricultural products. Developing countries have these countries as a template for modernizing their own national agricultural industries. In China, due to a huge population of over 1.3 billion and the demand for food has thus put huge pressure on agricultural production.

To meet this demand for food, the heavy use of pesticides and fertilizers has improved land productivity significantly. A survey by the Agricultural Ministry of China has shown that China's arable land accounts for only 7 % of the arable land in the world, but fertilizer use accounts more than 40 % of total world use. The average fertilizer use in China is about 400 kg/ha and is far higher than the recognized safe quantity of 225 kg/ha in developed countries.[2] At the same time, the use of pesticides in China's agriculture has reached about 1.3 million tons

[1] United Nations (2012).

[2] Ibid; see note 32, Chap. 1.

B. Qin, *Sustainable Development in Rural China*,
DOI 10.1007/978-3-662-46476-2_2

annually. However, about 1/3 of the pesticides can be absorbed by crops and the left has been left to pollute 93,000 km^2 of arable land.[3] The monitoring of vegetable wholesale market over 16 main cities in 2002 showed that the overall detection rate of pesticide was 20–60 %. It is far beyond the corresponding detection rate for developed countries.[4] Fertilizers and pesticides not only directly threaten human health, but also pollute the soil and water sources while damaging eco-systems. How to achieve sustainable agricultural development has become a huge issue for rural areas.

2.2 Research on the Sustainable Agriculture

2.2.1 *Agricultural Production and Land Management*

When examining the question of sustainable agriculture, it is essential that the productivity factor be investigated fully, for not only is the crop yield of agricultural land important with regards meeting the huge nationwide demand for food, but also the sale of produce is what provides agricultural workers with their main income. The kind of land system employed is crucial in determining how successful crop production actually is. After the founding of the People's Republic of China in 1949, the Chinese government passed "The Agrarian Reform Act"[5] which altered land ownership thus taking land away from the landowners and giving it directly to the farm workers. Since then, the Chinese central government has organized agricultural production into individual smaller groups of workers most commonly referred to as communes, collectives or cooperatives.

During the aforesaid period, the distribution of water to farmland was the responsibility of a group charged with conserving water that basically increased the land which had to be irrigated 1.25 times from 1952 to 1978.[6] As a result of this, agricultural production was improved and despite a 3-year period of natural disasters, it was possible to meet the minimal needs in the national demand for food. In addition to this, the communes were able to raise around RMB 80 billion which was the amount required to kick start China's industrialization program.[7]

However, the system of collectivization began to show a number of flaws in it as time went on. Under the communal system it was extremely hard to monitor quality and quantity of the work carried out on the farms. The equal distribution of land had severely dampened agricultural workers' enthusiasm for working on the land. This

---

[3]Ibid; see note 33, Chap. 1.

[4]Ibid; see note 34, Chap. 1.

[5]Chinese Central Government (1950).

[6]Lu (1992).

[7]Ibid; see note 6.

can be seen in the limited growth rate in China's agricultural sector from 1952 to 1978,[8] showing a mere 2.9 % (9). In respect of increase in workers' pay between 1952 and 1978, the net per capita annual income only came to RMB 2.88.[9]

As part of a post-'78 land reform program, the communal system was changed over to the family contract responsibility system. Since then, it has been identified as a cornerstone in Chinese economic policy towards the countryside. Under the contract system, the principle of stated owned farmland gives way to the principle of farming families owning their land. Based on this system,[10] farmers are encouraged to sell surplus produce on the open market and a small percent can be turned over to the state, thus they are able to keep the rest of the profit for themselves. Therefore this new system managed to change the productive unit from a collectivized one to a family run model.

Under the new system, workers developed a greater enthusiasm for their jobs. This yielded more favorable results in China's agricultural production. Farming families had been given the opportunity to decide what control they had over the kind of crops they produced. Therefore greater yielding crops produced an improvement in income. Naturally, the freedom to sell excess produce at the market made it possible to increase their income. This also had a knock on effect on the industrial sector as well. It has been proven that after the adoption of the family based system, agricultural output rose 4.8–7.7 %.[11] Agricultural growth rose in output from RMB 140 billion in 1978 to RMB 800 billion in '90, an increase of 4.73 times, and during the same period, net per capita income increased 4.3 times in rural China.[12]

However it must be said that due to increased industrialization, the system is no longer suitable for rural communities in China. First of all, small scale farming is causing an obstacle to large scale modernization. Under the family based system, it is difficult to mechanize agricultural production. After China's membership of the WTO, China lost its advantage in the agricultural sector as traditional methods are labor intensive and tend to push the price of agricultural produce up. Of course, China faces competition from countries that already use wide scale mechanized production in agriculture.[13]

Another point is that the present system puts workers in a weak position when doing business on the free market. The variance in pricing for a single product is enormous. As a result, agricultural workers must content themselves with whatever the going rate for a single product may be on the market. Furthermore, it is impossible for individual farmers to conduct the kinds of consumer surveys that

[8]Ma (2011).

[9]Ibid; see note 42, Chap. 1.

[10]In fact, no or only a verbal agreement existed.

[11]Ibid; see note 43, Chap. 1.

[12]Ibid; see note 42, Chap. 1.

[13]Tan (2008).

some of the big producers use in order to meet the demands of the market. In fact, the lack of such information has led to a lot of waste in the agricultural sector.

Finally, the family based system hinders the market allocation of land resources which are not being fully utilized. As the land provides not only job security, but also steady income, welfare and pension security, it is not hard to see why a farm worker would not give up the right to tenancy of the land. Even though a great number of agricultural workers have migrated to the towns and cities, they would rather leave the land abandoned than give up tenure. There again, for those workers who wish to expand their business and develop into large scale producers, the existing system is even more of a hindrance to them. Some researchers and experts in this field such as Wu[14] and Zhou[15] have studied the alternatives. It must be stated, however, that for agriculture to maintain any credible sustainability, agricultural pollution must be avoided. Agricultural development is so closely tied up with maintaining the natural balance with the ecosystem. Modern agricultural methods make a considerable impact on the environment. Sustainable agriculture should ensure supply as well as protect the environment. The big question remains how can sustainable agricultural development be maintained in China's rural areas?

2.2.2 *Agricultural Pollution and the Sustainable Agriculture*

To meet this demand for food, the heavy use of pesticides and fertilizers has improved land productivity significantly. Fertilizers and pesticides not only directly threaten human health, but also pollute the soil and water sources while damaging eco-systems. A survey by the Agricultural Ministry of China has shown that China's arable land accounts for only 7 % of the arable land in the world, but fertilizer use accounts more than 40 % of total world use. The average fertilizer use in China is about 400 kg/ha and is far higher than the recognized safe quantity of 225 kg/ha in developed countries.[16] At the same time, the use of pesticides in China's agriculture has reached about 1.3 million tons annually. However, about 1/3 of the pesticides can be absorbed by crops and the left has been left to pollute 93,000 km^2 of arable land.[17] The monitoring of vegetable wholesale market over 16 main cities in 2002 showed that the overall detection rate of pesticide was 20–60 %. It is far beyond the corresponding detection rate for developed countries.[18] Fertilizers and pesticides not only directly threaten human health, but also pollute the soil and water sources while damaging eco-systems. How to achieve sustainable agricultural development has become a huge issue for rural areas.

---

[14] Wu (2006).

[15] Zhou (2004).

[16] Ibid; see note 32, Chap. 1.

[17] Ibid; see note 33, Chap. 1.

[18] Ibid; see note 34, Chap. 1.

"Silent Spring"[19] is recognized as the first work which advocated the sustainable path to development. The author investigated the condition of the excessive use of DDT and other pesticides and herbicides in the United States and pointed out that the use of pesticides irreparably causes harm to the human body and can even the entire ecosystem. This work made the world reflect on conventional means to obtain economic growth and promoted some deep thinking on the subject of sustainable development. The idea of sustainable agriculture has gradually been taken seriously.

In order to resolve pollution problems caused by the agricultural production, many industrialized countries have attempted to adopt new farming methods instead of the conventional agricultural production methods. There has emerged a variety of concepts and theories including those of Organic Agriculture, Biological Agriculture, Natural Agriculture, Ecological Agriculture and Sustainable Agriculture. They all reflect a strong desire for and exploration of sustainable agricultural development even if the concept and content are different.

In 1981, a large-scale organic agriculture symposium co-sponsored by the American Society of Agronomy (ASA), Crop Science and Soil Science Society was held in Atlanta. As a result of the symposium, the document of "Organic Farming: Current Technology and Its Role in a Sustainable Agriculture"[20] was published. Almost at the same time, groundwater pollution had become increasingly serious in the European Community[21] member states due to intensive use of fertilizers and pesticides. The Netherlands, Belgium and other member states started the Eco-regional Methodology Approach to promote cross-ecological, economic and sociological interdisciplinary approaches in addressing this problem.[22]

2.3 Sino-Japan Comparative Analysis of Eco-agriculture

Japan began to notice the need for sustainable agricultural methods in addressing this problem earlier than the U.S. and EU countries. In the private sector, Fukuoka (1913–2008) is believed to be the founder of natural farming methods and began to try it as early as 1937. Since then he researched it further by using the opportunity to work at the Kochi Agricultural Center and form a system. Based on the natural faming theory, he summarized that it is important to make full use of the inherent strength of the soil. Unnecessary fertilization had caused land degradation and had weakened the regeneration. Through years of research and practice, he published "Natural Farming – One Straw Revolution"[23] and "Return

[19]Ibid; see note 10, Chap. 1.

[20]Bezdicek (1984).

[21]EC is the predecessor of EU.

[22]Dangjian Yanjiu (Research on the Establishment of Communist Party) (2011).

[23]Fukuoka (1983).

to Nature"[24] and some other books on the subject. The Japanese government has taken sustainable agricultural development very seriously. This county introduced and implemented a number of effective policies and formed its own system. Japan's agricultural operation model is family-based and small-scale and so it is the same as China's in this respect.

In Japan, in addition to sustainable agriculture, organic agriculture and ecological agriculture are two labels which are widely used. In this chapter, I shall define what sustainable agriculture is and determine the direction of agricultural development as well as ecological agriculture and organic farming in their various and specific forms. The Japanese government takes the promotion of sustainable agriculture very seriously and considers organic agriculture as an effective method to sustainability.

2.3.1 Technological and Financial Support

As the Japan's authority on agriculture, the Ministry of Agriculture, Forestry and Fishery (MAFF) are in charge of implementing technical guidance in order to promote the use of organic farming methods. According to the local climate and soil characteristics, MAFF set up an agricultural technology center on a countrywide basis which engaged in scientific and technological experiments into organic agriculture. It was intended to popularize scientific and technological achievements among local farmers and agricultural organizations. At the same time, the agricultural technology center held agricultural training seminars at regular intervals for Japan Agriculture (JA) to train farmers who hoped to engage in organic agricultural production.

In addition, MAFF configure assigned promoters of organic farming in Agricultural Bureau to each province. The promoters not only carried out promotional activities but also provided farmers and organizations with pro-bono consulting services. Furthermore, MAFF set up organic agricultural demonstration sites throughout the country and provided financial subsidies for it. In 2009, the number of demonstration sites was up to 47 and financial subsidies reached 4.5 billion yen.[25]

Besides the central Government, local governments actively promote and support local sustainable agriculture as well. Each prefecture has made a plan to promote organic agriculture. Because organic products are exclusively certified by the MAFF, local governments have adopted the term "Ecological Agriculture". The contents of "planning to promote organic agriculture" by local governments have a lot in common place with MAFF. In addition to detailed planning, each prefecture has substantial input in providing ecological agriculture support systems. Local governments support agricultural organizations and farmers by engaging them in

---

[24]Fukuoka (1984).

[25]MAFF (2011).

ecological agriculture by investing capital, providing training personnel, consulting services and other similar activities.

2.3.2 Support to Expanding Market

In addition to the promotion of production activities, the Japanese government also engaged in expanding understanding and recognition for organic products among consumer. In order to expand the market for ecological producers, as the country's authority responsible for eco-agriculture, MAFF pour great effort to expand understanding and recognition of organic products among consumers. MAFF advocated the promotion of the organic agricultural products through various channels and achieved encouraging results. The web-based survey carried out by MAFF in 2007 showed that the recognition rate of organic products had reached to 95 %, and nearly half of consumers wanted to buy organic products on a daily basis.[26] The only problem was to get consumers to accept the higher prices. Two thirds of consumers just had to accept up to more than a thirty percent increase in the price of organic goods.

For the local government, they always help eco-agricultural producers to expand the market directly. Firstly, the local governments have ensured governmental purchase of ecological agricultural products and so for example the cafeteria in all public schools should promote the sale of certificated products, in return, government agencies therefore become the consumers of eco-agricultural products. Secondly, local Governments may host trade fairs inviting eco-agriculture producers, agro-processing enterprises and supermarkets to help to establish a network. Meanwhile, governments provide places for eco-farmers and organizations to sell their goods and develop markets to promote the sale of eco-agricultural products. In order to increase local eco-agricultural propaganda, governments have allocated space to promote eco-friendly agriculture on local government's website. In addition, the governments send professionals to go to the major agricultural markets to introduce the locally produced eco-friendly products.

To enhance consumer confidence, agricultural products are always labeled with a record including the information of using fertilizer and pesticide even with information about the use of fertilizers and pesticides can even display the producer's photos and contact details. In this way, governments provide support in the three areas of production, circulation of products and the solution to difficulties and problems encountered by the ecological farmers and organizations. This has made it possible for the farmers to focus on agricultural production.

[26]Ibid; see note 25.

2.3.3 Eco-agricultural Certification

Certification is considered an effective method by which the sale of ecological agricultural goods can be promoted. On the one hand, it can increase consumer's recognition of organic products; on the other hand, it can encourage producers to improve the product quality. Eco-agriculture certification in Japan can be divided into central governmental and local governmental level as well. At the central government level, MAFF designated the corporate association of Japanese Agricultural Standards (JAS) to develop a standard certification system for agricultural products. Compliant with the most strictly certificated standards, under the premise of there being no GM foods, agricultural products have to be cultivated for a period of 3 years without the use of chemical fertilizers and pesticides until the JAS certification could be obtained.

The JAS certification system is very strict and has a certified number which remains small. On the contrary, local governments have thus established eco-certification systems at several levels. I selected two case of Fukui and Chiba due to each of them is near to big city in t eastern and western Japan thus has a great eco-agriculture market. In Fukui Prefecture, an eco-agricultural certification system includes specially cultivated crops and eco-farmer crops. Such agricultural products have reduced the use of chemical fertilizers and pesticides by more than 50 % greater than the percent used by conventional cultivation methods in order to pass Fukui Prefecture's "specially cultivated crops" certification process. "Eco-farmer" is the other certification which refers to the agricultural worker or organizations that are engaging in soil management and have reduced the use of chemical fertilizers and pesticides by more than 20 % compared to conventional cultivation methods.[27] Besides Fukui Prefecture, other local governments have adopted ecological farming certification systems as well. There are lots of common contents among them, for instance, Chiba's ecological farming certification system.[28]

The division of eco-agricultural certification has stipulated clear targets for farmers to work towards and for organizations interested in and committed to the development of ecological agriculture. For the farmers who want to get JAS certification, they should engage in cultivating crops without the use of chemical fertilizers, pesticides and transgenic seeds for over 3 years. Because JAS certification is so strict and the success of obtaining it is so great uncertain, the farmer should take quite a high risk. However, under the division of certification by stages, they can apply for local "eco-farmer crops certification" by firstly reducing 20 % their use of chemical fertilizers and pesticides which is relatively simple. After they have proven to be success, they can try to apply for "specially cultivated crops certification" and engage in reducing the use of chemical fertilizer and pesticide by 50 %.

[27]Fukui prefecture (2011).

[28]China State Council Information Office (2011).

The two stages of success will give the eco-agricultural engagers great confidence to meet the challenges of JAS certification. Through these efforts by stages, it is possible for JAS to increase the number of farmers who engage in eco-friendly forms of agriculture. But in fact, it is still very difficult to obtain JAS certification and so the number has not increased. In contrast, to this "eco-farmer" and "special cultivated certification" have made great progress. For instance, application for JAS certification remains at 137.[29] However in Chiba Prefecture, applications for eco-farmer crops certification went from 164 in 2002 to 3116 in 2008, and applications for specially cultivated crops went up from 568 in 2002 to 6190 in 2008.[30]

2.4 Concluding Remarks

Through my research into the Japanese government's strategies to develop ecological agriculture, I have found that in China the government should support sustainable agricultural development in the following way:

Firstly, as family-based land operations and small-scale agriculture is the general condition that we find rural China, the family-based land operation needs financial and technical support by being provided with the initial capital investment and technology development that is required in order to set eco-friendly agricultural methods in place. Anyway, the farmers and agricultural business organizations themselves are the fundamental carriers of development of ecological agriculture. The small-scale farmers and organizations don't have enough financial power to carry out the research, so it is necessary to get support from the governments. Without government's support in science and technology development, high ecological awareness and enthusiasm cannot be converted into production actions. Therefore, the Chinese government should increase financial input and technological support in its promotion of sustainable agricultural production.

Secondly, as market demand is a prerequisite for production, the government should conduct investigations into and manage to expansion of the eco-agricultural market. Using the market mechanism, the government can help agricultural farmers to deliver and carry out sustainable agricultural production and distribution through a planned and structured sales system. In order to format and promote eco agricultural markets, Chinese government should encourage governmental purchase of ecological agricultural products and ensure that the cafeteria of all public schools, government agencies become the consumers of eco-agricultural products. To expand the market of ecological agricultural products, government should hold conference among farmers, agro-processing companies and supermarket sales organizations, and through a series of activities to format the network of production and marketing. In short, because demand is the source of production, Chinese government should

[29]Ibid; see note 25.

[30]Chiba Prefecture Agricultural (2011).

provide marketable support to ecological farmers and organizations and make it possible for them to focus on production.

In addition, because certification is one of effective means to expand ecological agricultural market, Chinese government should also establish a multi-level certification system to help sustainable agriculture. It can not only improve the product's quality but also help formation of the brand and thus contribute to high price. In China, because national income has been increasing and consumers are become taking food security serious, there is no doubt that the market of ecological agricultural products will be gradually expanded. Reasonably high price and gradually expanded market of eco-agricultural production will undoubtedly increase the income of residents in rural China. Therefore, Chinese government should also establish a multi-level certification system. The development of ecological agriculture will no doubly become an effective means of poverty alleviation in rural areas.

References

Bezdicek DF (1984) Organic farming: current technology and its role in a sustainable agriculture (ASA special publication). American Society of Agronomy, Madison

Chiba Prefecture Agricultural Cooperative (2011). http://www.ja-chibamirai.or.jp/. Accessed 1 July 2011

China State Council Information Office (2011). http://www.gov.cn/zwgk/2007-12/26/content_844159.htm. Accessed 11 July 2011

Chinese Central Government. (1950) Xinhua net. http://news.xinhuanet.com/ziliao/2004-12/14/content_2331893.htm. Accessed 23 Nov 2011

Dangjian Yanjiu (Research on the Establishment of Communist Party) (2011) The practices of modern agricultural in Netherlands and Belgium. http://www.zgdjyj.com/Default.aspx?tabid=99&ArticleId=266. Accessed 26 Feb 2011

Fukui prefecture (2011) http://www.pref.fukui.jp/. Accessed 3 June 2011

Fukuoka M (1983) Natural farming: one straw revolution. Syunjyusya Press, Tokyo

Fukuoka M (1984) Return to nature. Syunjyusya Press, Tokyo

Lu W (1992) Research on agricultural development and the achievements of reform in rural China, Chinese Communist Party History Research. Chinese Academy of Social Sciences Institute of Party History. http://www.cass.net.cn/zhuanti/y_party/yc/yc_f/yc_f_074.htm. Accessed 13 May 2010

Ma Y (2011) On the advantages and disadvantages of the household contract responsibility system, Modern Marketing, No. 8

MAFF (2011) National conference on the promotion of organic agriculture. http://www.maff.go.jp/j/seisan/kankyo/yuuki/y_kaigi.html. Accessed 8 Sept 2011

Tan YW (2008) Policy research on improving the subsidies to China's cotton industry. China Economic Publishing House, Beijing, p 152

United Nations (2012) Population size and growth. http://www.un.org/zh/development/population/growth.shtml. Accessed 7 Jan 2012

Wu SL (2006) Agricultural development in rural China: systematic analysis and empirical study. Shanghai Financial University Press, Shanghai, pp 134–151

Zhou QR (2004) Property and institutional change: research on the experience of China's economic reform. Peking University Press, Beijing, pp 38–72

Chapter 3
Research on Industrial Pollution in Rural China: Seeking a Solution to the Pollution Incurred to the Farmers

3.1 Industrial Pollution in Rural China

Recently, with the development of industrialization in rural China, the industrial zones and industrial parks have been emerging in rural areas throughout the countryside.[1] At the same time, because many high-polluting enterprises have been emerging in rural regions, the rural environmental pollution incidents have appeared frequently. Especially water pollution is serious. More than 50 % of China's major waterways are heavily polluted.[2] In rural China, 360 million people are using unsafe drinking water poisoned by pollution. Among them, 190 million rural residents are drinking water in those toxic substances exceeding the required safety standards.[3]

Water pollution in rural areas is mainly from three sources; the first is due to excessive application of fertilizer and pesticides in agricultural production: second, due to lack of treatment facilities and management of life garbage led to water pollution as well. The third reason is from industrial production in rural regions.[4] However, China's water is mostly polluted by industrial wastewater and sewage, especially, the industrial factor contributed to 63.8 % of the total.[5]

In fact, Chinese central government has been improving its attention shown to environmental problems gradually. For instance, in the Environmental institutional settings, 30 years ago, China just established an Environmental Protection Office under the Construction Committee. The office became National Environmental Protection Agency in 1987. In 2009, it was upgraded to Ministry of Environmental

---

[1]Chen and Huang (2007).

[2]Dexter Roberts (2003).

[3]Chinese Central Government, Water Pollution Prevention Law, Article 35. 1984.5.11.

[4]Chinese Central Government, Water Pollution Prevention Law, Article 36. 1984.5.11.

[5]Meng and Ma (2002).

B. Qin, *Sustainable Development in Rural China*,
DOI 10.1007/978-3-662-46476-2_3

Table 3.1 Key Chinese environmental law

The Air Pollution Prevention and Control Law (1987, amended in 2000)
The Clean Production Promotion Law (2002)
The Energy Conservation Law (1997)
The Environmental Impact Assessment Law (2002)
The Environmental Noise Pollution Control Law (1996)
The Solid Waste Pollution Prevention Law (1995)
The Marine Environmental Protection Law (1982, amended in 1999)
The Radioactive Pollution Prevention and Control Law (2003)
The Water Pollution Prevention and Control Law (1984, amended in 1996)
The Wildlife Protection Law (1988)

Source: Richard and Zhang (2003)

Protection led directly by the State Council.[6] The fact that the institution in charge of environmental protection went from an office to a ministry reflects the Chinese government's deepening concern for environmental protection. The question is why rural environmental pollution problems have not been effectively curbed even though the government has been paying more attention to environmental protection?

Some scholars such as Ma[7] believe that China's financial investment for environmental governance is not enough. However, during the tenth Five-Year Plan from 2001, the central government pledged to spend 1.3 % of its GDP on the environmental sector, with the majority of this money focused on pollution treatment and prevention. Some scholars such as Wu[8] believe that lack of environmental law is one of the reasons for China's environmental degradation. However, through the legislation, we can say that China is not lacking in environmental law. China introduced the industrial three wastes discharge trial standards in 1973. As the first environmental protection law, the Environmental Protection Law (Trial) was passed in 1979. And since the economic reform, the environmental protection system has been formed gradually in China (Table 3.1).

Some scholars such as Eric[9] believe that China's environmental legislation has been developing continuously; China's environmental problems cannot be resolved because law enforcement has been weak. In order to find the real cause of environmental pollution in rural areas, I empirically investigated a chemical company and analyzed it in this chapter.

[6] Fang (2009).

[7] Ma (2007).

[8] Wu (2004).

[9] Eric and Jennifer (2004).

3.2 Field Survey of a Chemical Plant's Water Pollution

3.2.1 Overview of the Survey

Through the survey of a chemical company, this chapter seeks to: (1) realize the station of industrial water pollution in rural areas; (2) find out the reasons for long-term unresolved pollution; (3) give reasonable policy suggestions for pollution control in rural China. I have carried out a survey on a chemical company in July, 2010. I interviewed the factory managers and factory workers, a number of villagers living near the factory, the local Environmental Protection Agency officers, local hospital experts and the chemical industry expert. This chapter adopted two survey methods including: (1) field observation, to understand the actual plant pollution; (2) individual interviews. Even the information from only one company is not enough; I hope to realize the reasons of industrial pollution with the analysis of a typical company.

In this chapter, the object of the survey is Hua Chemical Company. It was started in 1996, after 10 years of establishment, it developed into a big enterprise which has 1,600 employees and 510 million RMB sales in 2006, and became one of the corporate major tax payers for local government. As the biggest maker of DSD acid, the output reached 30,000 tons per annual. Most products are exported to developed countries such as European nations, America and Japan. It meets over 60 % of the global DSD acid needs. DSD acid is the commodity name of 4,4′-diamino-2,2′-stilbenedisulfonic acid and a necessary dye and pigment intermediary. The raw materials are nitro-toluene, sulfuric acid and caustic soda.[10]

Carrying the questions of how productions of DSD acid damage the environment, I consulted the expert of chemical department of Zhanjiang Normal University. The production of DSD acid concludes mainly three steps of nitro toluene sulfur nation, oxidation and reduction. The materials are highly corrosive and the products usually contain toxic contaminants such as formaldehyde, benzene and heavy metals. Due to the water media oxidation process in production, there are problems in the process which are unstable, difficult to control, have poor response and cause extremely serious pollution. Especially in water pollution, about 30 tons of high polluted water will be discharged during the production of 1 ton of DSD acid. However, it is a big issue in resolving this pollution and there isn't any real technical solution. So production of DSD acid is prohibited in most developed countries, and they imported the products from developing countries where there is a lack of environmental protection.

For these reasons, this product has a large market in the world and therefore it is very profitable. However, in fact, the profit is on the loss of agricultural production and the health of local residents. Based on the information of Hua chemical group, the production of DSD acid reaches 30,000 tons and 30 ton waste water for each

[10]HP (2011).

product, this company discharges 1,300 ton waste water every day. However, the waste of other products is not included. It is easy to image how this company damages the local environment.

3.2.2 *Field Observation*

Hua Chemical plant is located in Dongguang County, Hebei Province and here is the North China Plain, one of the most populated areas in China. The North China Plain is China's second widest plain, and the population reaches 450 million and accounts for about 1/5 of the total. It has an area of about 310,000 km^2 and is one of the largest agricultural regions. The production of grain and cotton accounts for 18.4 % and 40.0 % of China's total output. The coastal area is also one of the most important sources of sea salt.[11]

Even though many rivers flow through the North China Plain such as the Yellow River, the Huai River and the Hai River, but the anural of water is less. The water volume in this region is only 335 m^3 per capita annually and less than 1/6 of the countrywide average. So groundwater has become an important pillar for sustainable economic and social development in this region. Currently, most cities in the North China Plain are relying heavily on groundwater and the share of groundwater has reached more than 70 % of the total water supply.

The factory is about 3 km far from southern Dongguang County and next to State Road 104. There are several villages around the factory and a kindergarten is next to it just separated by a wall. There is a river between the factory and State Road. The river flow is dark and it smells bad. Behind the factory, there is a river about 1 km far behind the factory and it flows to the Xuanhui River, which flows across hundreds of villages and into the Bohai Sea. There are several ponds next to the factories and the water is dark and smells bad as well. A villager working in the farm told me that these are sewage storage tanks from the Hua chemical company. There are pipelines buried under the soil, and the chemical factory emits waste water on rainy days or at night. The waste water has seriously polluted the rivers and groundwater.

3.2.3 *Interviews the Related Staffs*

3.2.3.1 Villager Survey

The river used to fill the water source originally, one villager told me, but since the chemical company came and emitted waste water into the river, the river has been polluted and it has caused a reduction in crop yields. According to villagers,

[11]Ministry of Land and Resources of China (2012).

Hua chemical group has caused pollution problems on many occasions. One of the villagers seemed to be frightened, and he said: "The sewage plant was horrible." They irrigated land once in 1999 with the river polluted by Hua Chemical Company, and crops died. The villagers made a video of that, and then petitioned the local government to resolve the problem. Finally, Hua Chemical Company was forced to make compensation of 70,000 RMB. The residents of Xu village told me that the plant's pollution damages not only the crops but also livestock.

As local drinking water totally depends on the groundwater, the company's chemical production damaged the health of local people. According to a local villager, Hua chemical plant has been polluting the water for many years since its foundation in 1996. Initially, because they did not know the harm that the chemical pollution caused, they just noted that the air had become thick and the river water dark. However, 2 years later, they found that not only was the river totally black and choking, but even 30 m-deep underground, it had became yellow and smelt bad. They had to drink the water, and sometimes waited for 2 days until the color had changed. A middle-aged farmer said that shallow wells could not be used as drinking water at all. For that reason, Hua chemical company dug several 300 m deep wells for the nearby villages. Even the water is from so deep ground, it is slightly brown in color.

The Xiaoxing villagers said that since Hua Chemical Company moved to their village and polluted the water source, the number of cancer patients has increased. According to the villagers' statistics, several dozens cancer patients have been discovered in the nearby villages in recent years, and many of them have died. Many residents of nearby villages had developed cancer. The villagers said that if you go to Cangzhou hospital, you will see that most cancer patients are from Dongguang County. Besides cancer, in Dongguang County, the incidence of femoral necrosis is much higher than the surrounding counties. In addition, other strange diseases have begun to appear, especially the occurrence of several children being born with deformities.

The villagers have complained and petitioned repeatedly, but the government has not resolved the problems. The pollution problems here have been also exposed by several media outlets. But this has not been of any help in solving the problem. So there is no change in the problem of pollution here. Villagers have decided to resolve the problems by themselves and attempted to prevent the pollution from business activities on several occasions. So there have been several conflicts between the corporation and the residents. As a result, County Public Security Bureau (Police Office) has ended up mediating and detaining the organizers. Therefore, the villagers have lost confidence and hope for a solution to their problem. They only told me what they had reported to the Deputy Prime Minister, which is to say the problem would have to be resolved or, it wouldn't work.

3.2.3.2 Hospital Survey

After that, carrying the prevalence of cancer, I visited Dongguang County Hospital[12] and Cangzhou People's Hospital.[13] It seemed that they were reluctant to submit specific figures, but through interviews of several hospital staff, I came to a conclusion. In Dongguang County, the incidence rates of various types of cancer has soared in recent years and the cancer incidence rates of this county are far more than in surrounding counties.

3.2.3.3 Survey on the Local EPA

Concerning the pollution from Hua chemical company, I visited Dongguang County Environmental Protection Agency and interviewed the executive officer. He said that the company did have pollution problems. As a petition had been received on many occasions, EPA and county government communicated and coordinated with the Hua chemical corporation. Eventually, the company bought sewage treatment equipment and solved the pollution problems. The enterprise treats the sewage and then pumps it to a sewage treatment plant to be treated. So there is no problem with the final discharged water.

The executive officer told me that because pollution control targets for this company are lower than the requirements of the Government, the State Environmental Protection Administration and the Provincial Environmental Protection Bureau have not found any problems. When I asked about the problem of the high incidence of cancer and other diseases among local residents, the executive officer showed me an investigation report from 2007 conducted by the Department of Environmental Protection of Hebei Province. It clearly stated that the incidence of cancer in Dongguang County is lower than the national average.

Through the investigation conducted by the EPA of Dongguang County, It can be sure that the Hua chemical company does not have pollution problems. Even assuming that the problems were resolved after the sewage treatment plant had been established, nevertheless the county's sewage treatment plant was only built in 2005, which is 10 years after the foundation of the chemical plant. In other words, the pollution problems have existed for 10 years. According to experts, 30 tons of highly polluted waste water will be emitted due to production of each ton of DSD acid. The company produces 15,000 tons DSD acid annually, and 10 years of production of DSD acid has already had discharged 4.5 million tons of highly polluted wastewater. This means that the company has already caused extremely serious damage to local water resources. The other problem is whether the pollution problems have been really resolved or not. Bearing this in mind, I triedto visit the

[12]Address: Eastern Fuqian street, Dongguang County, Cangzhou City, Hebei Province.

[13]Address: North Main Street 20, Canal Zone, Cangzhou City, Hebei Province.

company. However, I was refused an interview when I contacted the company and described the investigation I was carrying out. Therefore I decided to interview some employees to understand the situation more clearly.

3.2.3.4 Employee Survey

According to one of the plant's workers who asked that I not disclose his name, the factory purchased the devices for sewage treatment. However, these facilities will be operated only during checks by the Environmental Protection Agency or their correspondents. After they leave, the operation will be stopped because running costs are too high. In addition, one worker said that the factory planned to establish branch plants in other regions, but failed several times to invest because of pollution problems.

3.3 Survey Analysis

Through on-site observation and survey of relevant personnel, we can see that the chemical plant caused great harm to the local environment, especially the water source. However, why could not the problem of pollution be effectively resolved given the indisputable evidence? Based on the survey, we can conclude that the reasons are as follows:

3.3.1 Environmental Law and Enforcement

The environment protection law is vague and this makes it difficult for law enforcement. However, China already has a groundwater protection law system, including "Water Law", "Water Pollution Prevention Law," "Water and Soil Conservation Law" and "City Water Regulations". These laws provide a number of measures for the protection of groundwater resources. However, these laws generally focus on how to ensure the supply of groundwater, not the protection of groundwater quality.

At the same time, although some laws make provision for water pollution, these provisions are too vague and limited. For example, to prevent groundwater pollution, the "Water Pollution Prevention Law" has made the following provisions: Prohibition of the emission of waste containing toxic pollutants, of sewage including pathogens and other waste from the seepage of wells, seepage pits, fissures and caverns,[14] and the prohibition of the transportation orstorage of wastewater

[14]Ibid; see note 3.

containing toxic pollutants, including pathogens of sewage and other waste using ditches and ponds without anti-leakage measures.[15] However, the law has not clearly defined groundwater quality.

3.3.2 *Beyond the Bureaucracy*

In China, since the officials are often keen to develop the local economy and even seek personal gains, it is common that the local government protects the polluters. The local governments in rural regions have often viewed the economic development as its priority task. In order to develop the local economy, they diffuse investment widely but pay less attention to its environmental impact. The environmental management of local government is not strict. On the one hand, they are environmentally aware about business site-settlement. On the other hand, if enterprises have pollution problems, the local government often condones it and even takes the attitude that the enforcement of environmental protection should not hinder the pace of economic development.

At the same time, because heavy polluting enterprises often manage to forge relationships with the local officials by giving shares or positions to them, in return, officials provide information concerning environmental checks from the SEPA to polluting enterprises. Once the inspection teams have arrived, the enterprises manage to address the pollution problems and meet the standards. However, when the inspectors leave, they often continue to fail to meet the pollution standards which would put heavy cost burden on them. Some local governments even refuse to cooperate with the State Environmental Protection Administration (SEPA) in punishing factories which caused pollution. Furthermore, local officials often pressure the EPB to ignore pollution from enterprises.

In addition, because environmental officials are subject to local government economics, the central Ministry of Environmental Protection law enforcement cannot be carried through to the final stage of the process. In China, the State Environmental Protection Administration (SEPA) is the chief architect of national environment protection standards, regulations and laws. It is responsible for supervising the implementation of national laws at the national level. However, implementation is overwhelmingly a local issue. Local environmental protection officials are responsible for the enforcement of the environmental regulation of a regional level. However, the EPB holds a dual allegiance to the SEPA and the local government. On one hand, they are branches of the SEPA. On the other hand, they depend on the local government for financial support. Their salaries and office maintenance fees are all from local fiscal expenditure. So even if the Environmental Protection Bureau attempt to enforce environmental regulations, they have to consider how it influences the local officials.

[15] Ibid; see note 4.

3.3.3 *Transfer of Industrial Pollution*

In the context of economic globalization, transnational pollution has become one of the factors of environmental damage in rural China. In the context of industrialization and abstract investment in rural areas, many companies have transferred their production bases to the rural industrial zone including many heavily polluting enterprises. One of the very important sources of pollution enterprises in rural areas is from heavily populated cities. In order to reduce pollution pressures, the central cities began to evacuate the seriously polluting plants. This makes the seriously polluting enterprises spread out from urban areas to rural regions, thus many rural areas have begun to become industrial areas with their own polluting problems.[16]

Another source of pollution from enterprises is that of the developed industrial countries.[17] With economic globalization and international production division, cross-border pollution is increasing. On the one hand, because the regulation of industrial pollution in the advanced industrial countries is very strict, many heavily polluting enterprises in these countries have selected to transfer operations to developing countries where the environmental regulation is much looser. On the other hand, because the production process of some products has caused very serious environmental pollution, it is prohibited in developed countries. So enterprises in developed countries have taken to importing their goods from developing countries.

Due to a huge market and industry clusters, China is one of the most important locations for these enterprises. This has contributed to the country's environmental problems and has made industrial pollution in developing countries heavier. In China, especially in eastern rural areas, the rural economy is relatively developed and local industrial enterprises have the production capacity to meet the demands of the international market. So many chemical companies have accepted the heavy production of pollution and have become a base of high-pollution production. In the context of trade liberalization and economic globalization, environmental problems are spreading worldwide. In other words, economic globalization and the pollution globalization are part of the same process. Therefore, under the background of globalization, researching China's sustainable development should deal with environmental problems on the global scale.

3.4 Concluding Remarks

In order to achieve sustainable development in rural China, solving the environmental pollution problem is one of the government's most important tasks. First of all, China should change the economic growth mode by changing the "treatment after

---

[16]Zhang (2007).

[17]Jin (1997).

pollution" model and establishing pollution prevention and a control system. The regional economic development plane should be based on the regional environmental agency carrying out its capacity to prevent pollution from urban to rural areas. As for the examination of site-settle industrial production, construction of high-pollution projects should be firmly banned in rural areas. Especially with regards to the pollution that takes a long time to resolve such as groundwater, the examination should be particularly careful. In 2002, the Environmental Impact Assessment Law was promulgated and became the basis for environmental for project site selection. In the future, local governments should comply strictly in accordance with law enforcement.

Secondly, improvement of legislation and enforcement is the foundation of pollution prevention and control. When we review the global situation, we can see that the industrialized countries have also experienced the process of environmental pollution resolution. These countries finally have achieved the condition of economic development needed for appropriate environmental protection because they have established a sound legal system. So China should improve its environmental protection laws, especially in the context of economic globalization, legislation, pollution indicators and should gradually move closer to the position of move developed countries.

Thirdly, at the same time as improving the legislation, China should establish strict enforcement and effective monitoring systems. Environmental protection and company management should be put together, and the law enforcement can then ensure that the production cannot go beyond environmental regulations. SEPA should strengthen environmental law enforcement on industrial enterprises in rural areas and carry out remedial action for industrial pollution. Furthermore, the environmental protection agency should monitor the emissions from industrial enterprises and crack down the illegal emissions. In particular, the SEPA should take some nationwide action to resolve pollution problems which strongly affected residents in rural regions. For the enterprises in which such pollution cannot be resolved technically such as the Hua Chemical Company mentioned previously, its production activities should be resolutely banned. If local government officials condone their production activities, the producers and connivance of government officials should be severely punished.

Fourthly, the central government should establish an effective mechanism to urge local governments to pay more attention to environmental protection. Because local officials have played such a decisive role in the decision-making on economic development and environmental protection, enforcement significantly depends on them. The central government should change the evaluation content of local government performance and view environmental protection as its most important task. Especially for the rural governors, industrial pollution control and environmental protection indicators should be an important part of the appraisal index system and the necessary assessment of rural cadres. On the other hand, the central government should establish a mechanism for environmental protection target responsibility and accountability encouraging the local officials to engage in environmental protection work.

Finally, the Government should increase investment on environmental protection in rural areas. On the one hand, it is important to study and formulate environmental prevention and control planning according to the characteristics of rural pollution, while at the same time, encouraging pollution prevention technology development in rural areas and thus improving environmental protection capability. On the other hand, because environmental officials are subject to local government for financial aid, the central Ministry of Environmental Protection law enforcement cannot be carried out through to its final stages to uncooperative local governments. To ensure that environmental enforcement is independent, the central government should establish a special salary fund for an environmental protection system by making the officials economically independent from local governments. In addition, the government in rural areas should improve the public education of cadres and residents to actively participate in environmental protection.

References

Chen JG, Huang QH (2007) Report of China's industrialization process (1995–2005). Social Science Press, Beijing, p 2

Dexter Roberts (2003) The greening of China. Business Week, 2 Oct 2003, p 53

Eric Z, Jennifer LT (2004) Beyond the bureaucracy. East Gate, New York, p 98

Fang CL (2009) The urbanization process and environmental protection in China. Science Press, Beijing, p 47

HP of Hua Chemical Group (2011) http://www.hua-chem.com/product/hgindex.htm. Accessed 2 June 2011

Jin LSh (1997) The international and urban–rural transfer of environmental pollution. China Environ Sci 17(4):335–339

Ma YL (2007) Research on risk-sharing mechanism of environmental investment. Northwest Univ J 20(3):105–109

Meng X, Ma FS (2002) Groundwater resources and environment in China. Earthquake Press, Beijing, p 65

Ministry of Land and Resources of China (2012) http://www.mlr.gov.cn/xwdt/jrxw/201110/t20111027_1017726.htm. Accessed 7 Jan 2012

Richard JF, Zhang H (2003) Reaching out to the rule of law: China's continuing efforts to develop an effective environmental law regime. William and Mary Bill of Rights J 11:47

Wu J (2004) On the reform of environmental management system in China. Environ Protect J 3:13–15

Zhang ZQ (2007) Thinking of laws on rural industrial wastewater pollution control. J Anhui Agric Sci 35(3):84–85

Chapter 4
Sustainable Energy Supply for the Development of Rural China: Research on Wind Power

4.1 The Increasing Energy Demand in Rural China

Energy is the material basis of human society's survival and development. Our human civilization has reached an unprecedented height today. This was based on great energy consumption during the early and middle stages of industrialization especially in developed countries. Energy is extremely important for economic development such as modern agriculture, industry and transportation. At the same time, it is used in our daily lives such as for cooking, heating our houses and service facilities which also depend on energy heavily. Therefore, we can say that no modern society or civilization would be possible without a sustainable energy supply.

In modern China, since the establishment of the new country in 1949, energy consumption has grown steadily. In particular, after the adoption of a reform policy and an opening up of the economy to the rest of the world, China's economy has developed rapidly. At the same time, as Fig. 4.1 showed, the energy consumption has also increased sharply. Especially in coastal regions, because the scale of investment in fixed assets has expanded and the energy supply cannot keep up with the increasing demand, the shortage of the energy has become an acute problem as it which puts a halt to local economic development.[1]

The gap between domestic supply and demand has made China change from an oil importer in 1993, and its dependency on energy imports is expected to grow over the next two decades. Some analysts estimate that China will need to import about 60 % of its oil and at least 30 % of its natural gas by 2020.[2]

[1]Jiang (2008).

[2]Erica (2006).

B. Qin, *Sustainable Development in Rural China*,
DOI 10.1007/978-3-662-46476-2_4

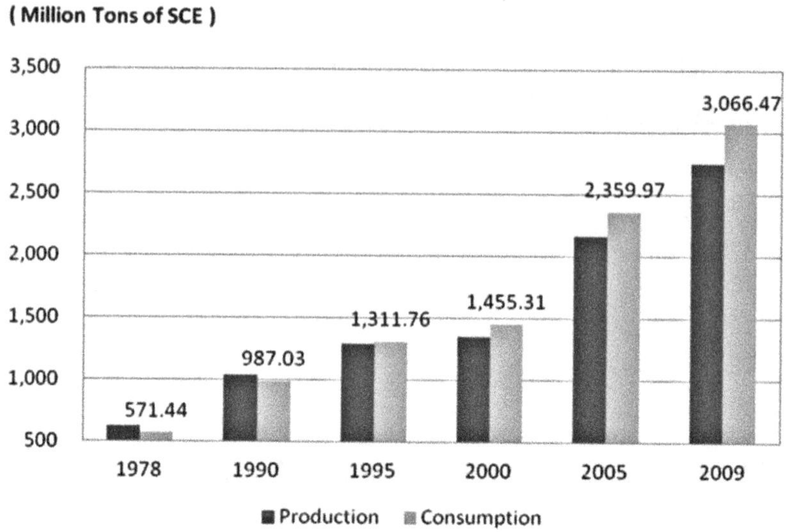

Fig. 4.1 Total energy production and consumption of China (Source: NBSC, China statistical yearbook, 2000, 2005, 2010)

Table 4.1 The number and capacity of agricultural machine in China

Year	1978	1990	2000	2009
Total power (million kW)	11,749	28,707	52,573	87,496
Large tractors (unit)	557,358	813,521	974,547	3,515,757
Small tractors (unit)	1,373,000	6,981,000	12,643,696	17,509,031
Diesel engines (unit)	2,657,000	4,111,000	6,881,174	9,249,167

Source: NBSC, China statistical yearbook, 1983–2010

On the contrary, the energy supply in rural China is extremely poor. For rural residents, electricity consumption per capita is only 316 KWh.[3] Furthermore, nearly 500 million people are living without electric facilities.[4] Resolving energy security is a necessary requirement for improving a rural resident's livelihood. However, on the other hand, with the Modernization of agricultural production in rural areas, using agricultural machines has gradually spread as Table 4.1 showed, and this has also increased the demand for energy.

At the same time, 30 years of economy growth have improved rural residents' living standards. As Table 4.2 showed, the penetration of household electrical appliances has gradually increased. This is another factor for the energy demand increasing. Moreover, industrialization in rural China has also developed rapidly. Industrial zones and industrial parks have been emerging throughout the country-

[3]Xinhua Net (2011).

[4]National Development and Reform Commission (2011).

Table 4.2 Housing machine per 100 rural households

Year	1990	1995	2000	2005	2008	2009
Washing machine	9.12	16.9	28.58	40.2	49.11	53.14
Refrigerator	1.22	5.15	12.31	20.1	30.19	37.11
Air conditioner	–	0.18	1.32	6.4	9.82	12.23
Color TV set	4.72	16.92	48.74	84.08	99.22	108.94
Motorcycle	0.89	4.91	21.94	40.7	52.45	56.64

Source: NBSC, China statistical yearbook, 1990–2010

side. Rural industrialization has brought great challenges to China's energy security. Energy security should be guaranteed not only for improving rural residents' living standards, but also for the development of modern agriculture and the promotion of industrialization in rural areas.

Fossil fuels are China's main energy source. In particular, the proportion of coal accounted for 70 % of China's total energy consumption.[5] Burning fossil fuels will affect the sustainable development of society. On the one hand, burning coal causes several types of emissions such as sulfur dioxide (SO_2) and nitrogen oxides (NO_x) which are two major air pollutants. On the other hand, non-renewable fossil fuel resources will eventually dry up. The traditional fossil energy-based energy consumption patterns cannot be changed totally in the short term. As the core source of energy, it may still be used for some decades to come. However, we should consider sustainable energy security. Firstly, fossil energy's future supply and pollution control technology issues should be consider. The second is the alternative energy development before the depletion of fossil fuels.

Renewable energy is one of the alternative energy sources to replace traditional fossil fuels. The expanding use of renewable energy sources such as wind, solar and biomass is considered one of the most sustainable development paths. Especially, there is a China's vast territory and very rich wind resources in rural China. So developing wind power is one of the important choices in energy security and sustainable development path for rural regions.

China's wind power electricity-generation is developing extremely quickly. As will be shown in the following section, total installed capacity in China has risen to rank fourth in the world at the moment. In addition, the new installed investment in 2009 was over the United States and ranked number one in the world. This is considered to be the result of powerful stimulus policies.

In this chapter, firstly, I will state the energy structure in China, and point out that depending heavily on coal is not sustainable due to the prospect of supply capacity and air pollution. And then, I will make note of condition of the wind power development in China. Thirdly, I will research the China's policy about the wind power and classify this into three ways: energy policy, the creation of a wind energy market framework and the fiscal and financial support for wind power generation.

[5]See Fig. 4.3.

Moreover, to verify the roles of these policies, I have done a survey of a wind power company and have released the results for this chapter. Lastly, I will indicate what the Chinese sustainable energy strategy should be.

4.2 China's Energy Structure and the Prospect

With about 30 years of economic development, as Fig. 4.1 in 4.1 shown energy demand in China has been increasing rapidly. The total energy consumption in 1980 was about 600 million tons of SCE, and it increased to five times of that to 3,000 million tones of SCE in 2009. Firstly, China's spectacular economic growth is majorly responsible for its rising energy demand. With the process of industrialization, especially the development of heavy industries including power, steel, nonferrous metals, building materials, oil refining and the chemical industry. This has accounted for the majority of energy consumption and has promoted China's energy needs which continue to grow.[6]

Secondly, the huge population and increase in living standards have contributed to the rise in energy demand. Even though China has 1.36 billion people, energy consumption per capita is still very low compared to the developed countries. Figure 4.2 shows the comparison of energy consumption per capita. At the same time, the economic development of China has increased people's living standards. So, the increasing in the use of household electrical machines and vehicles has also led to the increase in energy demand.

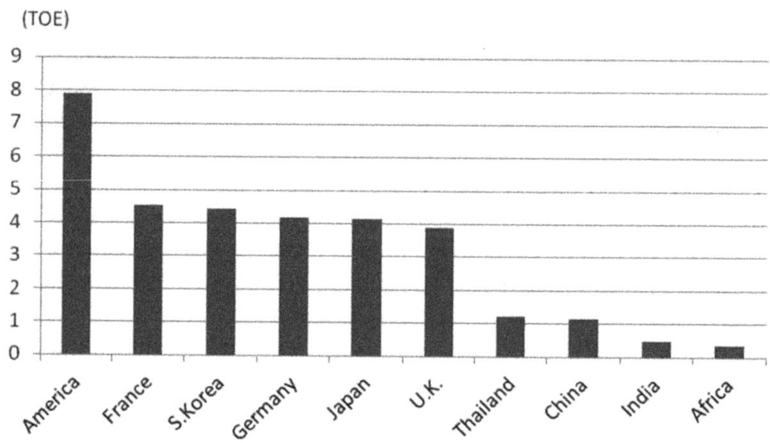

Fig. 4.2 Energy consumption per capita (2008) (Source: IEA 2009)

[6]International Energy Agency (IEA), (2009).

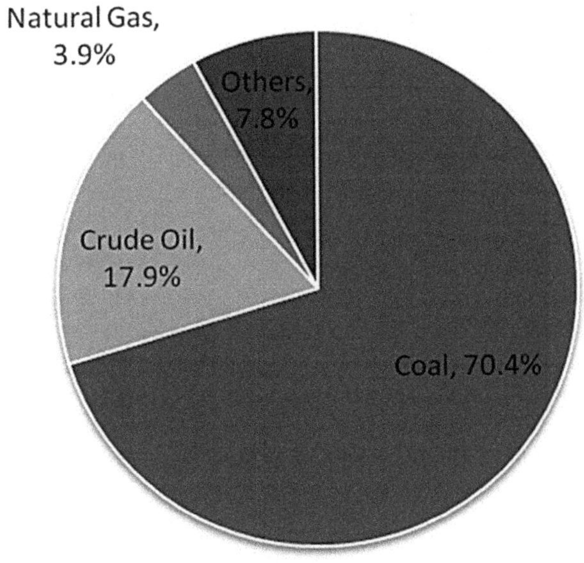

Fig. 4.3 Share of energy consumption in China (2009) (Source: NBSC, China statistical yearbook, 2010)

4.2.1 China's Energy Structure

China's energy consumption depends heavily on coal. As Fig. 4.3 showed, we can see that the share of coal consumption reached to over 70 % in 2009. Coal in China has not only been used for electric power generation, but also in the industrial sector mainly in the energy-intensive industry such as chemical, cement and steel. Besides, coal is also used directly in households for heating and cooking.

The use of coal is the major reason for air pollution and acid rain in China. Burning coal causes several types of emissions such as sulfur dioxide (SO_2) and nitrogen oxides (NO_x) which are two major air pollutants. China has been suffering from the damaging effects of acid rain and sulfur dioxide pollution. For instance, acid rain formed from SO_2 covers about 30 % of China's total land mass.[7] The emission of SO_2 is spread over 60 % of the whole country.[8] Table 4.3 shows SO_2 pollution in China's main cities.

To resolve these problems, China's central government has focused on the most polluted regions and has named it "Two Controlled Zones".[9] The government has adopted an acid control plan in the "National Economic and Social Development Tenth Five-Year Plan[10]" in 2001. Based on this plan, the central government would

[7]Wang and Li (2005).

[8]Eric and Jennifer (2005).

[9]To control acid rain and sulfur dioxide in these zones.

[10]Approved by the Fourth Meeting of the Ninth National People's Congress, 2001. 3.15. http://www.people.com.cn/GB/shizheng/16/20010318/419582.html, accessed on 2011.8.10.

Table 4.3 SO_2 Pollution in China (2005)

City	Days over 150 ug/m^3	Days over 230 ug/m^3
Beijing	68	272
Guangzhou	30	123
Shanghai	16	133
Shenyang	146	219
Xi An	71	273

Source: NBSC, China environmental statistical yearbook, 2006

reduce sulfur dioxide emissions in 2005 by 20 % to that of 2000.[11] At the same time, China will strive to make more than 80 % of the cities' dioxide air quality standards reach the secondary level in this area.[12] Because of the energy structure being based on coal it is the cause of the sulfur dioxide pollution, the government has therefore come up with a plan aimed at both the supply and demand sides in resolving this problem.

4.2.2 *The Bottleneck of New Coal-Using Technologies*

New coal-using technologies could reduce the sulfur emission. If innovative methods are successfully deployed, coal may largely define China 's energy path for at least the next few decades. On the contrary, without the new technologies, just burning coal directly is not a sustainable path for China. These technologies include Coal Washing (CW), Integrated Gasification Combined Cycle (IGCC), Coal-To-Liquids (CTL), and Underground Coal Gasification (UCG).[13]

4.2.2.1 Coal Washing (CW)

Coal washing can produce an average reduction in sulfur emission of 29 %.[14] In China, 18 % of coal was washed and it is assumed that this will grow to 50 % by 2030.[15] Even washed coal can generate less air pollution than the regular type; it is still greatly polluted by this method.

[11]National Economic and Social Development Tenth Five-Year Plan, Chapter 15, Section 2.

[12]Ibid; see note 25, Chap. 2.

[13]Richard (2009).

[14]United Nations (1994).

[15]Tu (2008).

4.2.2.2 Integrated Gasification Combined Cycle (IGCC)

One of the drawbacks of IGCC technology is that the gasification process would not work at more than 100 m above sea level. The other drawback is its high cost. The US Department of Energy has estimated a cost of $1,491 per kW.[16]

4.2.2.3 Underground Coal Gasification (UCG)

The basic process is conducted by a controlled underground burning of coal and then the combustible gas is collected on the ground surface. It can be used to recycle the old abandoned and high-sulfur, high-ash and high-gas coal mines. Because this technology is one of the alternative means of artificial coal mining, it can prevent mining accidents. Besides, UCG process purifies gas in a concentrated form and leaves the ash under the ground; it can therefore reduce air pollution. China initiated several UCG trials, of which 16 are still ongoing. However, technology is only practical if coal seams possess special properties. They should be between 100 and 600 m underground with a seam thickness of more than 5 m.[17] There should be minimal discontinuities in the seam, and no large water aquifers close by. The coal itself should have an ash content of less than 60 %.[18] Even though UCG is ongoing in China, technology applies to only a small portion of China's coal reserves and it is limited regards its wider deployment.

4.2.2.4 Coal-to-Liquids (GTL)

The basic process for CTL was developed at the beginning of the twentieth century and was used by Germany during the Second World War. In China, CTL technology was welcomed and invested on a large scale by Shenhua Group, which is the largest country-owned coal supplier. The key factor in the competition with regards CTL is the cost comparison with oil. The cost of CTL is about $67–82 per barrel depending on the price of coal and water.[19] Because of high projected costs, many proposed plants have been canceled in China. It is assumed that these projects could be started once the price of oil increases and coal has gained the through competition.[20]

Because China's energy supply depends heavily on coal, concern about the energy security are rooted in projections of its production capacity. Many scholars have researched the coal production forecast in China. Tao and Li, proved that

---

[16]Ibid; p. 137.

[17]Ibid; p. 138, see note 29, Chap. 2.

[18]Richard and Zhang (2003).

[19]Ibid; p. 133, see note 29, Chap. 2.

[20]Höök et al. (2008).

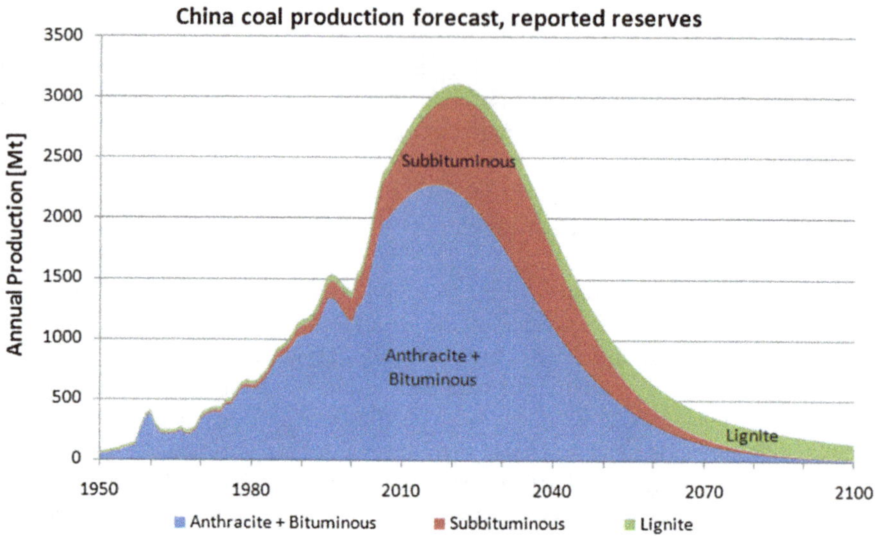

Fig. 4.4 China coal production forecast, reported reserve (Source: Höök et al. 2008)

the peak in China's production appears between 2025 and 2032. Their analysis was based on the Chinese government's official data applying Hubbert's analysis method.[21] Other research from Höök, et.al has suggested that the peak in China's production will be reached in about 2020.[22] As Fig. 4.4 shows, based the report, China's coal reserve may decrease sharply after the peak of 2020. So they have forecast that the changing trend in the future will be as follows. It will show that after the peak in 2020, coal production in China will decline steeply and it will cease finally in about 2100. The same scenario has been suggested by Zittel.[23]

Previous research has suggested that the peak in China's coal production would come between 2020 and 2030. The shift in 10 years is not important, but the problem is: what will replace coal as the main pillar of China's energy supply after production has peaked? Because of energy security issues and environmental protection concerns, China needs to gradually optimize its energy structure. This is especially true of the large-scale developments of renewable energy resources, which has recently become an extremely important strategy for sustainable development. Renewable energy including hydro, biomass, wind, solar, geothermal and ocean energy are important resources for sustainable development. Wind, biomass, solar power, and hydro have the most potential for use in technology

[21]Ibid; see note 32, Chap. 1.

[22]Ibid; see note 1, Chap. 3.

[23]Zittel (2007).

currently available. Wind power development is especially significant, and is rapidly spreading throughout the world.

4.3 Wind Power: The Sustainable Energy Supply for China

Based upon reasonable speculations, the installed electricity generating capacity in China will be about 1 billion kW in 2020. If calculated on the power structure and efficiency (383 g of standard coal/kWh) from 2002, China's coal consumption for power generation will be nearly 14 million tons, and the total coal consumption will exceed 30 million tons. At this time, China's coal mining will be reaching its limit, and over half of her oil will need to be imported.[24]

4.3.1 Wind Power Development in China

China is rich in wind resources. According to a wind resource assessment, the useful wind power in China is about 1 billion kW. However, because of the intermittent feature, the cost of wind electricity is higher than the traditional material electricity. Even though the technical level in China lags behind the U.S., and the grid connecting wind power only began in the 80s, electricity from wind generation is developing rapidly in China. The total installed capacity tripled from 4.2 MW in 1989 to 12.2 MW in 2008, ranking fourth in the world.[25] As in the U.S. and EU countries, wind power development is receiving strong Chinese government support.

In recent years, the use of wind power has been developing rapidly throughout the world. By 2009, the accumulative electricity-generating capacity of wind power reached 157,597 MW. In the United States and some EU countries, the newly installed wind power capacity has become second in importance only to natural gas. Furthermore, because of technological developments, wind power electricity has become more efficient and is being developed on a larger scale.

Clean energy utilizing in China developed rapidly in China. Table 4.4 show that the installed renewable energy capacity of China increased to 52.5 GW and almost the same as U.S. Especially in 2009; China invested 34.6 billion USD in clean energy projects and took the top spot for the first time, putting the U.S. which invested 18.6 billion into second place. We can see that clearly through Table 4.5. China has been building overseas markets, at the same time, working hard to meet domestic demand.

---

[24]Liang and Chuang (2005).

[25]Chinese Renewable Energy Industries Association (CREIA) (2010).

Table 4.4 Top five in renewable energy capacity (2009)

United States	53.4 GW
China	52.5 GW
Germany	36.2 GW
Spain	22.4 GW
India	16.5 GW

Source: The Pew charitable trusts, 2010

Table 4.5 Top five in clean energy investment (2009)

China	$34.6 billion
United States	$18.6 billion
United Kingdom	$11.2 billion
Spain	$10.4 billion
Brazil	$7.4 billion

Source: The Pew charitable trusts, 2010

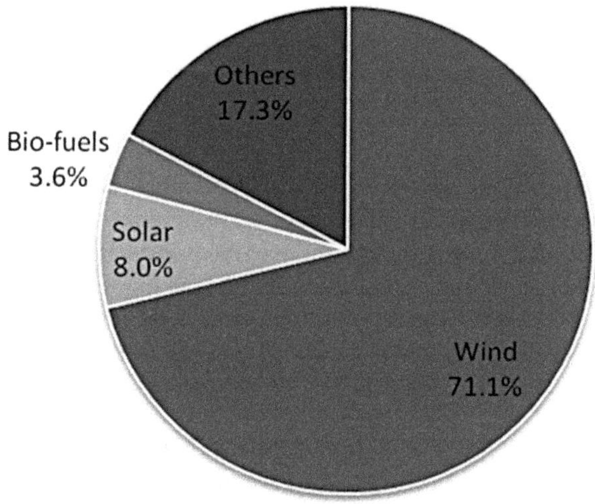

Fig. 4.5 Distribution of investment by sector in China (2005–2009) (Source: The Pew charitable trusts, 2010)

Within renewable energies, the development of wind power in China has increased most dramatically. China's rapid economic growth requires enormous amounts of energy. However, the traditional energy source in this county is coal. Burning coal not only causes serious air pollution, it is also not a sustainable energy resource. Pollution problems and increased environmental awareness from its citizens has led the Chinese government to shift its attention towards renewable energy resources. Next to hydroelectricity, wind power is now the second largest renewable energy resource in China. Figure 4.5 shows that from 2005 to 2009, the investment to wind power share the most and it is over 70 %.

China has a vast territory, a variety of terrains and an abundance of wind resources. It gives this country the physical conditions necessary to successfully develop wind power. Electricity produced from wind-power was first connected to the Chinese national grid in the mid-1980s. At that time, most of these initiatives

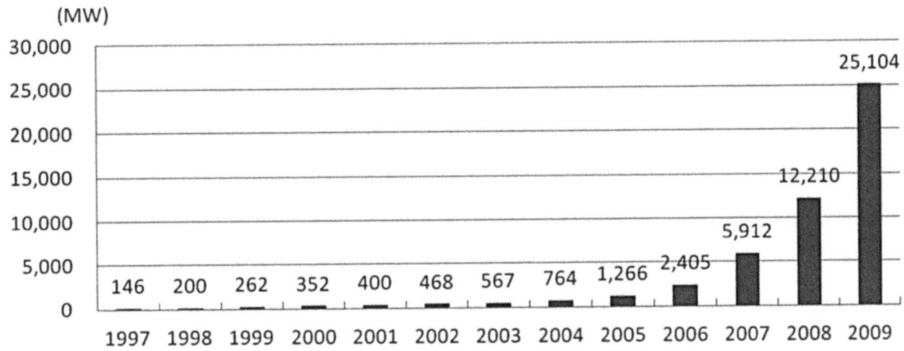

Fig. 4.6 The development of wind power capacity in China (Source: The wind power.net, 2010)

were small-scale wind power farms and were funded by foreign grants or low-interest loans. They were mainly used for scientific research or demonstration projects, but not for commercial operation. About 10 years later in the mid-1990s, China began to develop and generate wind power for its citizens. During this time, the government introduced the Cheng Feng Plan[26] and the Shuang Jia Project[27] to support wind power generation. Even though China was constrained by its low technological level, wind power was now being promoted. Figure 4.6 shows that wind power capacity was just 146 MW in 1997, and it increased five times to 764 MW in 2004. However, the newly installed electricity-generation machine's capacity was mostly lower than 100 MW.

From 2003 to 2007, the Chinese government introduced an Auction System for Wind Power Concession. Based on this system, government-owned, private and foreign companies have the same opportunities to do wind power projects. Thus wind power election-generating projects in China absorbed the investment not only from government but also from private and foreign companies, and the investment channel was expanded. Furthermore, after the mid-2000s, the government introduced several policies to promote renewable energy, resulting in a dramatic increase in wind-generated electricity. Figure 4.6 shows that the total installed capacity of wind power exceeded 25,000 MW in 2009. The three main energy policies that were introduced addressed energy policies, market frameworks and financial support. These three policies are discussed below.

[26]This plan was launched by the State Planning Commission of China in 1996. Its aim was to increase localization rate of large wind turbine manufacturing, detailed policies include encouraging technological research and cooperation with foreign enterprises.

[27]This project was launched by the State Economic and Trade Commission of China in 1994. To promote national industrial development, the commission selected 56 key industries and gave support to them. Wind power industry was in it.

4.3.2 Renewable Energy Policies in China

In December 2007, the Chinese government released a White Paper on energy titled China's Energy Conditions and Policies.[28] This eight-chapter paper firstly addressed the current status of China's energy conditions, its strategies, and its goals for energy development. Secondly, it stated that the government would promote energy conservation programs, improve energy supply capacity, and accelerate energy technology. Thirdly, the White Paper indicated that the government would reform the energy supply and demand system, and strengthen international cooperation in the energy field. Finally, but most importantly, with the aim to build a cleaner and more efficient energy society, the Chinese government addressed the importance of using renewable energy. The White Paper emphasized that China's national energy strategy should focus on efficiency, cleanliness and safety. As a result, clean energy development was written into the national energy development strategy.

In addition, in August, 2007, the Chinese government released a plan for the Medium and Long-term Development for Renewable Energy.[29] Firstly, this plan addressed the present situation of utilizing renewable energy. The progress of wind-powered electricity generation was highlighted and received a great deal of attention. In 2005, the total amount of renewable energy utilized in China was about 7.5 % of the total national primary energy consumption.

The medium and long-term objective of China is to raise the share of renewable energy to 10 % by 2010. By 2020, this share aims to be 15 %. This will be achieved by fully utilizing technologically mature and economically feasible renewable energy sources. Such sources include hydropower, biogas, solar thermal, geothermal, and wind power.

To realize this ambitious target of 15 %, the plan set priorities to develop the renewable energy sector. With respect to wind power generation, large-scale wind farms are viewed as the most viable option. As a result, the Chinese government aims to accelerate wind energy technology research and increase involvement by private industry.

The target date for wind power development was detailed in this plan. By 2010, the government aims to establish about thirty 100 MW-scale wind farms and increase the installed grid-connected wind capacity to 5 GW. By 2020, China aims to increase the wind capacity to 30 GW. In rich wind resource regions such as Hebei, Inner Mongolia, Jiangsu and Guangdong, the government will establish some large GW-level wind power facilities.

To support the development of renewable energy, this plan also addressed the necessity of additional policies and measures. Firstly, the report stated that the government needs to establish a stable market environment and improve sustainable demand. Secondly, because the cost of renewable electricity-generation

[28]Ibid; see note 28, Chap. 2, accessed on 2010 9.11.

[29]National Development and Reform Commission (2010).

is higher than traditional methods, cost-sharing policies are needed. Furthermore, the government should increase fiscal input and reduce taxes for renewable energy companies. Finally, the government should support additional technology research and increase involvement by private industry.

4.3.2.1 Market Framework Creation

In economics, the market is one of the effective means to allocate resources and distribute profit. As we know, China reformed its economy to marketable and achieved remarkable economic development. So marketwise is becoming the first selection to in development of wind power as well. In the current 10 years, the Chinese government has enacted many laws to build wind power market.

In February 2005, the Chinese Congress passed the Renewable Energy Law.[30] Article 14 of this law states that enterprises are required to either sign a grid-connection agreement with renewable electricity-generating companies, or to prove that they generate their own renewable energy. As a result, companies connected to the national energy grid are now required to consume part of their electricity needs from renewable energy sources.

At the same time, in the Medium and Long-Term Development Plan for Renewable Energy mentioned previously, the Government set a mandatory market share target for non-hydro renewable energy electricity generation. For energy producing enterprises which have a total installed power generation capacity exceeding 5 million kW, the share of non-hydro renewable electricity should exceed 1 %? by 2010. By 2020, this share will increase to over 3 %.

This plan established the demand for non-hydro renewable electricity in China. In other words, large electricity grid enterprises and big traditional electricity generating enterprises were pushed into the renewable electricity market. Both types of companies have a surplus of money and have the ability to either purchase or produce clean energy.

As the competent authority in China for electricity, the State Electricity Regulatory Commission released a file in July 2007 called the Monitoring Measures for Grid Enterprises to Fully Purchase Renewable Energy Power.[31] In this file, the Commission listed detailed monitoring measures. Firstly, the commission is to monitor whether grid companies purchase all electricity generated from renewable energy enterprises. Secondly, grid companies should construct an access grid for renewable electricity farms, and ensure that the grid-connection environment is completed on time. Thirdly, the grid access services will be monitored by the commission. Furthermore, electricity dispatching agencies should prioritize renewable energy electricity use whenever power output surpasses demand. Finally,

[30] The National Congress of China, Renewable Energy Law (2010).

[31] China State Electricity Regulatory Commission (2010).

grid companies should strictly follow their contracts to pay tariffs and subsidies to renewable energy generating enterprises in a timely manner.

Despite these measures, the price of wind power has remained high. To decrease the price of wind power electricity and expand large-scale wind power utilizing projects, the Chinese government adopted an auction system for wind power concession. A similar auction system was previously successful for natural gas and oil exploration projects belonging to the government, and for large infrastructure projects involving the mining of mineral resources. Under the concession agreement, private companies and the government share the interest and the risk.

In the wind power electricity-generation concession system, the main sectors are the government, electricity-generation companies and electricity grid companies. The policy can be divided into five sections. First, operation rights will be given to the company which is the winner of the concession auction. Second, the use of a certain percentage of Chinese equipment is required. Third, technical specifications and operational deadlines are required. Fourth, the electricity price for the first 30,000 kW is fixed on the auction price. After this, the price is determined by the market. Finally, electricity grid companies should purchase all the electricity generated by the generation companies. The responsibilities of the electricity-generation companies and the grid companies are therefore very clear. Generating companies take the responsibility of investment and management, and the profit and risk belongs to them. On the other hand, the grid companies have to purchase all of the electricity that is generated by the wind power electricity companies.

The Chinese government chose two wind power projects in Guangdong and Jiangsu provinces as the concession pilot projects. The auction commenced in March 2003, with nine companies attending including three foreign companies and one private company. As a result, the bid price in Jiangsu was 0.436 Yuan/kWh, and the other project in Guangdong was 0.501 Yuan/kWh. In 2004, the government successfully completed three more wind power electricity-generation concession pilot projects in Jiangsu, Inner Mongolia and Jilin province.[32]

The results show that the wind power concession project is a method which can not only decrease electricity prices of wind power in China, but can also absorb investment from private and foreign companies. The concession project succeeded in ending the high price history of wind power electricity. The total installed capacity of the five wind power projects reached 70 million kW, equals to that of the 20 years by then.

[32]Chinese Renewable Energy Industries Association (CREIA), Benchmark Electricity Price of Wind Power Announced by Chinese National Development and Reform Commission, http://www.creia.net/?Infors/detail/t/0/id/373.html, 2009.7.27.

4.3.2.2 Fiscal and Financial Support

In order to promote utilization of renewable energy and protect the environment, just relying on the market is not enough. One of the reasons is the cost of wind electricity is about two times high of material. There exist some other reasons including the low technological level and lack of human resources. Fiscal and financial support from government is very necessary.

Actually, The Renewable Energy Law, which was mentioned previously, contains much fiscal and financial support for renewable energy including wind power. First of all, the government establishes renewable energy development fund to support the programs such as, scientific and technological research, surveys, assessments of renewable energy resources. Secondly, the country-owned banks are urged to offer preferential loan with financial interest subsidy to renewable energy development programs. Thirdly, the government lists scientific and technical research in the utilization of renewable energy as hi-tech industrial development in the national programs. To promote technical advancement in utilization of renewable energy field and reduce the production cost, government gives technical support to renewable energy research programs. For long-view thinking, this law encourages the education of renewable energy technology and suggests the education authorities to incorporate the knowledge into general and occupational education curricula. Figure 4.7 shows the development of subsidy amount to wind power utilization.[33]

In areas not covered by the power grid, the Chinese government supports the construction of independent renewable power systems to provide power service for local production and living. The government encourages and supports various types of grid-connected renewable power generation as well.

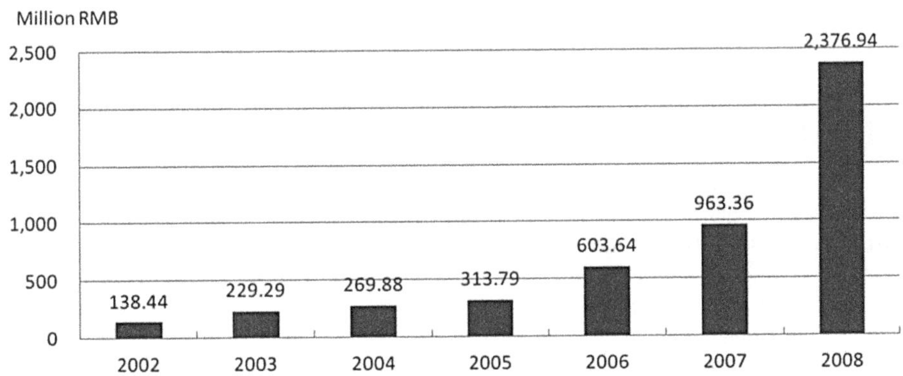

Fig. 4.7 The subsidy amount to wind power utilization (Source: CREIA 2009)

[33]Chinese Renewable Energy Industries Association (CREIA) (2009).

With respect to grid-connected renewable power, this law authorizes the State Council to set energy prices. The price for grid-connected power is also publicized. Price authorities of the State Council shall prepare specific methods for the exceed expenses that power grid enterprises purchase renewable power. Grid connection expenses paid by grid enterprises for the purchase of renewable power, and other reasonable expenses, may be included into the grid enterprise power transmission cost and retrieved from the selling price.

Renewable energy prices can be either government-fixed or government-guided. An example of a government-guided price system is the wind power electricity-generation concession bid. If the price is higher than the local desulfurization coal electricity price, the difference can be incorporated into the electricity price. With independent wind power systems, if operation and maintenance costs are higher than the average price of local desulfurization coal electricity, the difference can be charged to the final user. However, the grid companies should separate any additional price from their common account, and it should not be used for other purposes.

To support the development of wind power, the Chinese government established a special fund for renewable energy development and released interim management measures[34] in May, 2006. This fund aimed to support renewable energy development and utilization of scientific and technological research. In addition, it provided funds for renewable energy demonstration projects. Wind power was highlighted in the renewable energy list in Article 8.

The special funds for the above measures are provided as either grants or low interest loans. Grants are provided for non-profit and important public welfare projects. Low interest loans are mainly used for renewable energy development projects that are listed in the national renewable energy industrial development directory. The maximum repayment period is 3 years, with interest discounts up to a maximum of 3 %.

4.3.3 Field Survey of a Wind Power Enterprise

To verify the roles of these policies, I surveyed a wind power company in August 2010. I chose a representative wind power generation enterprise in Hebei Province. Because of the personal connection, I contacted the director of a power plant. I wanted to survey him and a senior manager of this company. Because the conduct to a tóp manager was not successful, I surveyed the director only. As the stuff in charge of the production, he learns more about the production and technical issues.

Even though his information about the whole company and is not comprehensive enough comparing with the top leadership, he knows the general situation of company. I plan to investigate deeply in the next step after the contacting to a senior

[34]China Ministry of Finance (2010).

manager successfully. Of course, the information from only one company is not enough. This chapter was written as a temporary research result and I hope that through the analysis of a typical example to draw some useful results.

The name of the company is Hebei Construction Investment New Energy Co., Ltd (HCINE), wholly owned by Hebei Construction Investment Group. It was established in 2006 and its main business is the development of wind, solar, nuclear and other new energy sources, providing clean power. The company's specific business plan is to take on new energy projects, development, survey, site selection and construction. Since its establishment, combining with regional development and construction condition of Hebei Province, HCINE has assessed wind power resources in fifteen counties, for instance, Zhangjiakou, Chengde, Cangzhou, Baoding, Tangshan, and Qinhuangdao. These regions cover almost all of the wind-rich areas. The company has signed agreements with these local governments and occupied up to 500 million kW of wind energy resources.

In 2006, this company established Wolong Mountain wind electric farm in Kangbao County. The farm has 40 machines and total installed capacity is 30,000 kW. This was the first farm HCINE has built and it was put into operation in November 2006. Next year, HCINE established the second wind farm Wuhuaping of Guyuan County. This farm has 36 machines and the total installed capacity was 30.6 thousand kW, it was put into operation in October the same year. In 2008, HCINE established two farms in Wei and Haixing Counties. There were 91 electric generators in the two farms and the total installed capacity up to 100,000 kW. These two farms were put into operation in June and July in the same year. In the next 2 years, HCINE established four farms and the biggest one has total installed capacity up to 200,000 kW. At present, eleven wind farms are in construction and the total installed capacity is about 100,000 kW. In addition, the projects in approval are 46 and the total capacity is over 2 million kW.

As an emerging new energy enterprise, HCINE aim at leading wind power. Presently, this company focuses on developing wind power in land region. At same time, to exploit maritime wind power in the future, HCINE established a hydrological and wind observation platform in Caofeidian. This marks that the company made a substantial progress.

In this chapter, I want to show that the local governments and enterprises made a great effort can connect with the sustainable development. We can know that HCINE was established in 2006 and from then the wind power electricity generation developed dramatically. About the problems in the field, the director told me that sometimes it is difficult to send electricity to the grid. He told me that is what he heard from the top managers. That indicated the necessity to survey the top managers and next I will do that.

4.4 Concluding Remarks

Wind power in China began developing in the 1980s. The progress attained since then reflects the special attention to wind power from the Chinese government. However, China's wind power industry currently faces many problems. Firstly, China's wind energy resource survey is not comprehensive enough. Abundant wind energy resources are extremely important when selecting a wind farm site. In China, the current distribution survey of wind resources is very vague, and it cannot meet the requirements of wind power companies to make a cost/benefit analysis. The Chinese government should refine the survey and provide more detailed wind distribution information to potential investors.

Secondly, the domestic technology level for manufacturing wind power equipment is not up to world standards. Wind turbine manufacturing technology needs to be nurtured so that wind power companies can further reduce costs. At present, the overall domestic wind power manufacturing level is lower than EU countries by at least 10 years.[35] China still does not have a large-scale wind turbine manufacturing capacity, and small turbine units produced are prone to failure. The Government therefore needs to increase its investment in Research and development to support various wind power enterprises.

Finally, financing channels for renewable energy are narrow. For wind power projects, the loan repayment period from domestic commercial banks is typically only 5–8 years. This is a shorter period compared to what is given to thermal power and hydropower projects, and the interest rate is not as preferential. On the contrary, wind R&D needs long-term investment so that the development of advanced wind power technology is not restricted. To this end, the Chinese government should establish a special financial institution to fund wind power research. Additional surveys and research can then be undertaken that will see China emerge as a world leader in wind power technology and generation.

References

China Ministry of Finance (2010) http://jjs.mof.gov.cn/zhengwuxinxi/zhengcefagui/200805/t20080523_34033.html. Accessed 14 Aug 2010

China State Electricity Regulatory Commission (2010) http://www.gov.cn/flfg/2007-08/01/content_702636.htm. Accessed 14 Sept 2010

Chinese Renewable Energy Industries Association (CREIA) (2009) http://www.creia.net/. Accessed 28 Nov 2011

Chinese Renewable Energy Industries Association (CREIA) (2010) http://www.creia.net/?Infors/detail/t/5/id/473.html. Accessed 9 July 2010

Eric Z, Jennifer LT (2005) China's quest for energy security. Rand Press, Arlington

Erica SD (2006) China's quest for energy security. Rand Press, Arlington, p 198

[35]Liu (2010).

Höök M, Werner Z, Schindler J, Kjell A (2008) Supply-driven forecast for the future global coal production, contribution to ASPO. www.tsl.uu.se/uhdsg/publications/Coalarticle.pdf. Accessed 1 Oct 2011

International Energy Agency (IEA) (2009) Energy balance of the world countries. http://www.iea.org/stats/index.asp. Accessed 12 Sept 2010

Jiang ZM (2008) Research on energy issues in China. AP Academic Press, London, p 132

Liang ZP, Chuang X (2005) Wind power concession: a new mechanism for guiding the wind power to large-scale development. China Construction News – Sun Energy.

Liu ChG (2010) Thinking about China's situation and development of wind power generation. http://www.lrn.cn/stratage/resposition/201009/t20100907_761618.htm. Accessed 20 Sept 2010

National Development and Reform Commission (2010) http://www.sdpc.gov.cn/zcfb/zcfbtz/2007tongzhi/t20070904_157352.htm. Accessed 13 Sept 2010

National Development and Reform Commission (2011) http://www.sdpc.gov.cn/gzdt/t20110711_422682.htm. Accessed 4 Oct 2011

Richard H (2009) Blackout: coal climate and the last energy crisis. New Society Publishers, Gabriola Island, p 129

Richard JF, Zhang H (2003) Reaching out to the rule of law: China's continuing efforts to develop an effective environmental law regime. William and Mary Bill of Rights J 11:569–602

The National Congress of China (2010) Renewable energy law. http://www.npc.gov.cn/huiyi/lfzt/kzsnyf/2009-08/23/content_1514455.htm. Accessed 5 Aug 2010

Tu JJ (2008) Modeling China's energy future: climate change air pollution and supply security. VDM Verlag, La Vergne, p 153

United Nations (1994) Energy and sustainable development: efficient utilization of energy resources: issue in the transfer of clean coal technologies to developing countries. Report of the Secretary-General, New York

Wang F, Li HF (2005) China's environment and the challenge of sustainable development. In: Kristen AD (ed) An east gate book. Armonk, New York, pp 76–77

Xinhua Net (2011) Energy agency: to solve the problem of last 500 million people without electricity power by 2015. http://news.xinhuanet.com/energy/2011-07/09/c_121645384.htm. Accessed 23 Dec 2011

Zittel WJ (2007) Coal: resources and future production, EWG-paper no.1.07. www.energywatchgroup.org/.../pdf/EWG_Report_Coal_10-07-2007ms.pdf. Accessed 1 Oct 2011

Chapter 5
The Study of Sustainable Labor Migration in China: Based on Lewis Two-Sector Model and a Field Survey

For researching sustainable development in rural China, labor migration is a very important area. This is because in the neoclassical theory of labor, it is a necessary element of economic development. At the same time, from China's actual perspective, the population and the amount of labor in rural China is huge. Compared with the limited arable land, there is a great surplus of labor force in rural areas and this is one of the direct reasons for that the ecological environment has been greatly damaged. Therefore, the transformations of the surplus of rural labor not only can ease rural environmental damaging, but also can promote the China's economic development.

Since Chinese has adopted the policy of open up and economic reforming, the surplus of labor has been migrating from the inner rural to coastal cities smoothly and contributed to the industrialization and the economic development of China. However, recently, some coastal areas have started to lack labor, especially in the Changjing River and Zhujiang River Deltas, which are the most developed regions and this has influenced local economical development. In this chapter, I will explore the sustainable labor migration with the Lewis's Two-Sector Model and analysis of a field survey.

5.1 The Surplus of Labor in Rural China and Sustainability

According to international statistical standards, the agricultural population refers to people who engaged in agricultural work and their economically independent families. However, in China, many people don't engage in agricultural production activities even though they belong to the agricultural population according to the statistics. So China's agricultural population which is accorded to the census data is always much larger than what it actually is in reality. Associated with this, the agricultural labor force has the same problem as well. Therefore, it is difficult to

B. Qin, *Sustainable Development in Rural China*,
DOI 10.1007/978-3-662-46476-2_5

Table 5.1 International comparison of agricultural population (2004)

	Agricultural population	
	(thousands)	(% of total)
World	2,599,791	41
Developed countries	82,592	6
Developing countries	2,517,199	49
Asia and the Pacific	1,872,666	55
Mainland China	846,304	66

Source: FAO, Population and Labor Force Indicators, 2005

Table 5.2 International Comparison of Agricultural Population (2004)

	Agriculture, value added (% of GDP)	Agricultural population (% of total)
China, Mainland	14.6	66
Philippines	14.5	37
Bolivia	14.9	42
Yemen	15.0	46
Azerbaijan	14.3	25

Source: FAO, Population and Labor Force Indicators, 2005

know the true surplus of labor in rural China. However, we can try to estimate it in international comparative methods on agricultural population, land size and agricultural output.[1]

We can estimate the surplus in rural labor with the agricultural population compared with the world. With Table 5.1, we can see that the share of agricultural population in developed courtiers is just 6 %. However, the same percentage in China is up to 66 % and is far higher than developed countries. And it even more than the world average of 41 %, the developing countries average of 49 % and the Asia & Pacific of 55 %. This means that the problem of surplus in labor of rural China is extremely serious.[2]

In China, as Table 5.2 shown, even where 66 % of the population is engaging in agriculture, the agricultural output only accounts for 14.6 % of the total GDP. This shows that productivity in the agricultural sector has been extremely low compared with that of the industrial sector. At almost the same level of agricultural share to GDP, there are four countries of Philippines, Bolivia, Yemen and Azerbaijan. Table 5.2 shows that two countries are a few lower than China and the other two are a few higher. However, as to the share of agricultural population in these countries, China is far higher than the other four countries. The average percentage of agricultural population in Philippines, Bolivia, Yemen and Azerbaijan is 37.5 and it is lower than that in China by 28.5. If we can consider that 28.5 % is the share of surplus in labor of total, the amount is up to 370.5 million.

[1]FAO (2004).

[2]Ibid; see note 35, Chap. 4.

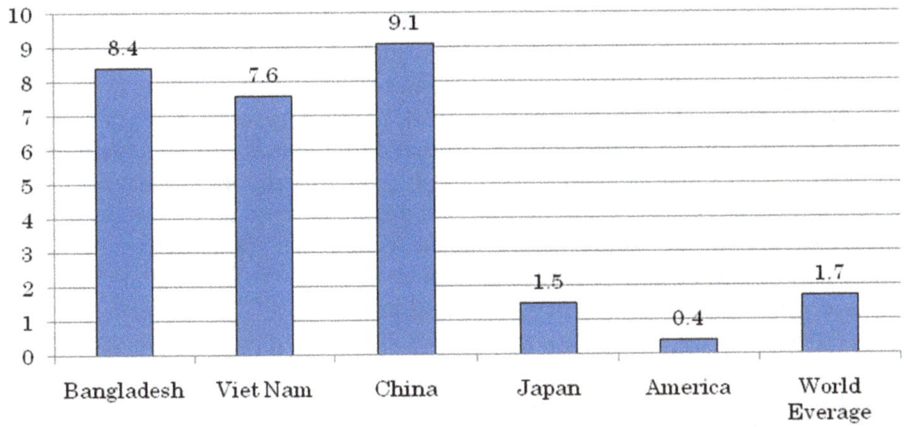

Fig. 5.1 Agricultural population per hectare of arable land (2004) (Source: FAO, Population and Labor Force Indicators, 2005)

Also, we can estimate the surplus in rural labor with the size of arable land. In 1958, China had 152 million ha of arable land and a population of 560 million. When it came to 1978, it had almost the same amount of arable land and a population of 832 million. Right now, the population increased to over 1.3 billion but the arable land has decreased somewhat due to construction and industrial use. From Fig. 5.1, we can see that the population per hectare of arable land in China is 9.1, even higher than Bangladesh and Vietnam. In contrast, the index is 1.5 for Japan, 0.4 for America and the world average for the world is 1.7. Based on the arable area and the level of agricultural productivity, rural China can accommodate about 200 million labor forces and 260 million is the surplus. Even when the development of township enterprises has absorbed half of it, there is about 130 million surplus of labor exists in rural China.[3]

In addition, as the mechanization in agricultural production, about 6 million of surplus of agricultural labor will generate annually in the next few years. Furthermore, about 10.8 million of the grown Children will become the workforce.[4] Therefore, rural China is enfacing tremendous pressure from the surplus in labor force.

Due to the surplus of labor force, the ecological environment in rural areas has been greatly damaged. Firstly, the area of forest and grassland has decreased. During the period from 1984 to 1991, the average of deforestation area has reached 558 thousands ha annually. Based on the data of the seventh national forest inventory, the forest coverage rate of China was 20.36 % in 2009, 10 % lower than the average

[3]Ibid; p. 87, see note 49, Chap. 1.

[4]Ibid, p. 88.

of the world.[5] Grassland area has gradually reduced and 1/5 of the total area of grassland in China have degraded.[6] Deforestation and the reduction of grassland have intensified the desertification and severed soil erosion. The area of decertified land has reached to 1.74 million km^2 by 2004, accounting for 18.12 % of China's entire territory. On the other hand, intensified deforestation and the reduction of grassland have caused 50 million tons loss of soil annually. In the last 50 years, 2.7 million ha arable lands have disappeared due to deforestation and reduction of grassland.[7] Furthermore, deforestation and the reduction of grassland have seriously harmed the biological diversity. There are about 4,600 kinds of plants and about 400 kinds of wild animals are in endangered status in China.[8]

In order to protect the ecological environment damaged by over-cultivation, the Chinese government has introduced the "Grain for Green Project".[9] Based on the policy, peasants has been subsidized by 1,575 Chinese Yuan in the Yangtze River area and 1,050 Chinese Yuan in the Yellow River basin each year, the subsidy period is 5 years for forestry and 2 years for grassland.[10] But these subsidies cannot compensate for the losses of reduction of agricultural production. To solve the problem of environmental destruction, government should reduce the rural population, and the transformation of surplus labor force is the only way out.

The transformations of the surplus of rural labor not only can ease rural environmental damaging, but also can promote the China's economic development. One the one hand, the migrant workers has provided the necessary labor for industrialization. On the other hand, it has increased the income of rural China. In the context of limited arable land and the agricultural technology, the reduction of rural population can increase the agricultural income per capita. Increase of transformation in surplus rural labor force by 1 % has contributed 2.28 % of increasing rural income per capita.[11] Therefore, the transfer of surplus of rural labor is an important issue for sustainable development in rural China. Since Chinese has adopted the policy of open up and economic reforming, the surplus of labor has been migrating from the inner rural to coastal cities smoothly and contributed to the industrialization and the economic development of China.

---

[5]Solid Wastes Management Center of Henan Province (2011).

[6]China International Ecological Production (2011).

[7]Forestry Administration of China (2005).

[8]Ibid; see note 4.

[9]It refers to converting the land for forestry and pasture.

[10]Xinhua Net (2011).

[11]Ibid; see note 49, Chap. 1, p. 112.

5.2 Labor Migration and Theories

5.2.1 Status of Labor Migration in China

China had been a typical dual society characterized by economic and institutional segmentation between rural and urban areas prior to 1978.[12] China had a special system of residency registration called the "Hukou" system. To control the migration and manage the productivity of all the country, the central government divided residency into that of rural and urban, based on where the populations were living. Peasants had to work in the commune and had no right to leave their hometown and live in the city. Even if they were to leave, it was impossible for them to exist independently due to the "food distribution system" as I shall describe later on.

On the other hand, the urban residents had to work in the same factory until they were asked to change occupation. The rural labor force were not allowed to work in off-farm activities or out of collective farms. Meanwhile, even though the economic development in coastal areas started to create employment opportunities for farmers. As a result, there was no migration in China during that period.

Since the Chinese government has adopted the heavy-industry development strategy, the government has relied on "price scissors" which were used against the agricultural sector and ensure the accumulation of capital into the heavy industrial segment. To guarantee agricultural productivity and maximize the capital flowing into the industrial sector, a food distribution system was adopted. In addition, the agricultural product free market was prohibited. Food was distributed to all the units based on the registered population number. That was another reason preventing the labor migration.

Even though both urban and rural residents were strictly controlled by the government; the insurance system was totally different. People working in the government-owned factories were covered by the social insurance programs including a pension system, medical insurance as well as occupational injury insurance. Besides, almost all of the central governors were selected from government-owned factories. On the other hand, agricultural workers had no other insurance except they could get medicine for free. In that era, wanted their children to enjoy the insurance and the opportunity to raise their social status, the rural residents managed to send their children to university, as that was the only means by which their children could change their rural residency status.

From 1978, the Chinese government started to reform the agricultural economy and initiated a "household responsibility system". The farm belonged to communes was divided into smaller areas which were then lent to small holders. Small holders got the right to plan productivity. At the same time, the agricultural product market had been opened and price functioning was given new life. That meant the more the peasants produced, the more they could earn.

---

[12]Meng and Bai (1988).

Another result of the reform is that the long-standing hidden surplus in the labor force was released. As has been mentioned previously, because the arable land per agricultural population is very low in China, there was huge number of surplus of labor which was hidden in the commune unit. Once the peasants got the right to decide productivity, labor migration occurred.

In the early stages of reform, many of the barriers which deterred agricultural labor from entering cities was not removed. The planners tended to control labor migration and established township enterprises to absorb the surplus agricultural labor. The government came up with a slogan of "leaving the land without leaving the village". Small scaled township enterprises eased the pressure on labor migration from rural area. Since township factories had the function of absorbing surplus agricultural labor and easing the pressure on labor migration, the government encouraged their development. However, at the end of 1980s, an overheated economy caused serious inflation. The central government tightened fiscal and monetary policies. China experienced its lowest economic growth rate. Certainly, the development of township enterprises was affected and many projects were halted as a result.

Entering into the 1990s, foreign investment helped to create a boom in China's economy. Many foreign factories emerged in the coastal areas and increased the huge demand for labor. The migration from the inner mainland to the coastal region occurred. From Fig. 5.2, we can see that migrant workers amounted to 61.99 in 1993. After that, migration continued and the amount of migrants rose to 136.49 in 1997. However, recently, some coastal areas have started to lack labor in China, and this has influenced local economical development. Some economists say that

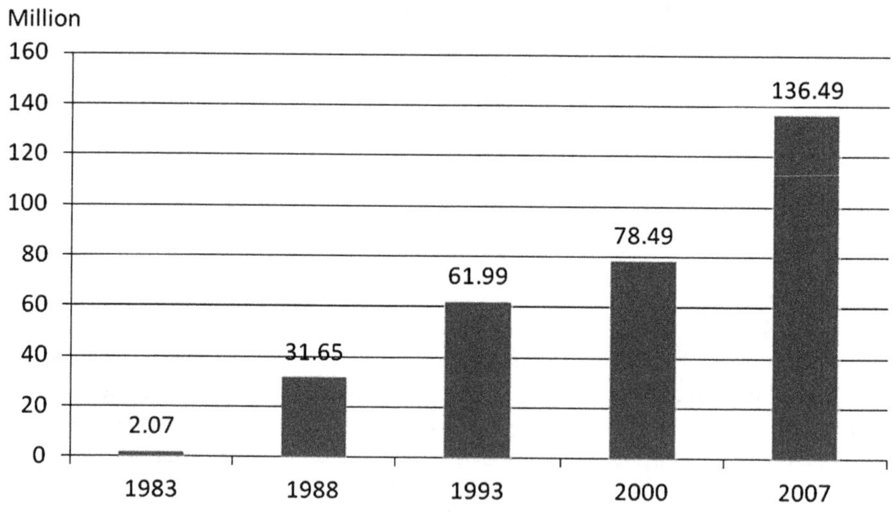

Fig. 5.2 The amount of migrant worker in China (Source: NBSC, China yearbook of rural households survey, 2008; NBSC, China statistical yearbook, 1983–2008)

China's development has reached the turning point as described in Lewis model. To analyze this problem, I will take a look at the Lewis Two-Sector Model in the next section.

5.2.2 *Reviews on Theories of Migration*

The earliest study related to the movement of people was carried out by Malthus. His research points out that the increase of food production cannot keep up with population growth and one of the consequences of this is that some of the population is forced to move out. Following Mlthus's study, many scholars have studied migration issues. E.G. Ravenstein systematically analyzed the rule of human migration from angle of distance, stage, direction and motives systematically. His studies suggest that people tend to move short distances and long-distance migration generally occurs in the larger industrial and commercial centers. He points out that the main direction of migration is from rural to urban areas, and among the many motives for migration, the economic factor is the main one, in other words, in order to improve their living conditions, people migrate from the low-income to the high-income areas.[13] On this basis, Lee has characterized the number and the features of migrants. He believes that migration is selective, and that the number of migrants moving to such areas has a very close relationship to the stage in economic development of these areas. Another result of research has shown that the characteristics of migrants generally are between the regions moved into and out.[14] Thomas, D.S. and Kuznetz, S. studied the relationship between economic growth and population migration. They point out that the added industrial demand for labor in a region's growth is met by the migration of workers into that region. Economic security is most important for migration and the opportunities of employment play a significant role.

This research into migration is vital. In particular, they point out that economic conditions are the main factors which affect migration and the general direction is from rural to urban areas and this seems to be the general. However, in a study of the transfer of rural, the most famous theory is Lewis's Two-Sector Model.[15] He believes that there exists two different types of sectors in a country's economy, one is the traditional agricultural sector and in which the productivity is low, the other is the industrial sector which has a more modern method of production and in which productivity is higher. In the agricultural sector, the marginal productivity of labor is close to zero and the labor force is surplus, the roll-out of labor does not affect labor productivity. In contrast, in the modern industrial sector, capital investment continues to expand the scale of production, and gradually increases the

[13] Ibid; see note 2, Chap. 1.

[14] Li (2004).

[15] Lewis (1954).

marginal productivity. Wages in the industrial sector are often higher than that of the agricultural sector and the gap is about 30 %. The pursuit of a higher income and the surplus of the agricultural labor force flow from the agricultural sector to the industrial sector until the wage gap between the two sectors have disappeared. The Lewis's two-sector theory is considered closely to the reality of China's situation, so it has become a useful method in researching labor problems in China.

5.3 Analysis on Lewis Model and a Field Survey

5.3.1 Lewis Two-Sector Model

The two-sector model of economic growth developed by Lewis, W. A.[16] states that the underdeveloped economy consists of two sectors: an agricultural sector and an industrial one. The traditional agricultural sector is typically characterized by low wages, an abundance of labor, and low productivity through a labor intensive production process. As a result, the agricultural sector has a number of farm workers that are not contributing to agricultural output due to the low marginal productivity. In contrast, the modern manufacturing sector is defined by higher wage rates than the agricultural sector, higher marginal productivity, and an initial demand for more workers.

This group of farmers who are not producing any output can be termed as surplus labor since this cohort could be moved to another sector without affecting agricultural output. The term surplus labor here is not being used in the Marxist context but only refers to the unproductive workers in the agricultural sector. Therefore, due to the wage differential between the agricultural and manufacturing sectors, workers will tend to transfer from the agricultural to the manufacturing sector over time to reap its benefits.

Over time as this transition continues to take place and investment results in increases in the capital stock, the marginal productivity of workers in the manufacturing sector will be driven up by capital formation and driven down by additional workers entering the manufacturing sector. The end result of this transitional process is that the agricultural wage equals that of the manufacturing sector, this means that the agricultural marginal product of labor equals the manufacturing marginal product of labor, and no migration takes place between the two sectors. In the next section, I will analyze the reality of the situation in China based on Lewis Model.

[16]Ibid; see note 12, Chap. 4.

5.3.2 A Labor Survey in a Coastal Factory

To realize the labor migration and the reasons for the labor shortage in coastal areas, I have decided to carry out a survey in a coastal city. For this purpose, the questionnaire among workers is a good method. The contents of the questionnaire have been designed to survey the following: (1) the motivations of migration from the rural to the urban; (2) the working and living conditions of migrant workers, especially their inconveniencies; (3) the sense of belonging to the cities; (4) awareness and motivation to return back to their hometown.

I have selected Shenzhen as the survey place because it is a typical city. As a special economic zone, here is the first pilot city of economic reform in China. During just 30 years, Shenzhen has developed from a small town to a huge modern city. Meanwhile, because here is the first pilot city for economic opening-up, wage levels are significantly higher than the inner mainland areas and this has absorbed a great number of rural surplus labors. The city's workforce has increased significantly from 130,000 in 1979 to 6.92 million in 2010.[17]

The survey has been carried out in electric company F in July 2011. 300 pieces of questionnaire were been distributed and 270 have been recovered; the recovery rate was 90 %. The contents of questionnaire include personal and family status, the motivation of migration, working and living conditions, and the awareness of returning back to their hometowns. The form of answer was multiple-choice questions, and some of the questions have free narrative column. I analyzed the questionnaires in cross-statistics mainly. Of course, the information from only one company and 300 workers is not enough. With the research result and the analysis of a typical example, I expect to discover the reasons for labor shortage in coastal areas and explore the sustainable labor migration in China.

F is a Taiwanese Company founded in 1974. As a branch plant, F was found in 1988 in Shenzhen mainly producing components for computers and other electric goods. Initially, it hired just about 100 employees. After then, it developed rapidly, expanding and eventually become largest electronics manufacturer in the world, and the number of employees has reached to about 240,000. As a foreign-funded enterprise which employs a large number of migrant workers from inner China, the manager has agreed to cooperate with me to carry out the survey.

Of 300 questionnaires, 270 were recovered and the recovery rate was 90 %. Table 5.3 shows the overview condition of the employee questionnaire. The gender of respondents was composed from 103 male and 167 female. Agricultural (Hukou) residence accounts for 67.4 %, and this indicates that most of the migrant workers in F Company are agricultural resident. As to the classification of their origins, 81.4 % of the workers are from the villages and towns. On the contrary, only 3 % of them are from cities. In addition, the workers are mainly from eight provinces of Hubei, Hunan, Jiangxi, Henan, Shanxi, Guangxi, Anhui and Liaoning. Especially the share

[17] Shenzhen Statistical Yearbook (2010).

Table 5.3 The overall status of all responders in company F

		Number	Percentage
Gender	Male	103	38.1
	Female	167	61.9
Residence	Agricultural	182	67.4
	Non-agricultural	88	32.6
Hometown	Village	151	55.9
	Town	69	25.5
	County	42	15.6
	City	8	3.0
Province	Hubei	100	37.0
	Hunan	61	22.6
	Jiangxi	29	10.7
	Henan	25	9.3
	Shanxi	20	7.4
	Guangxi	11	3.7
	Anhui	10	3.7
	Liaoning	8	3.0

of workers from the three nearby provinces of Hubei, Hunan and Jiangxi is up to 70 %. This may shows that relying on relatives and friends is one of the main means of migration.

5.3.3 Analysis on Lewis Model

In the Lewis Two-Sector Model, if the supply of labor from agricultural region decreases, then the average wage in the industrial sector will go up. And the migration from the agricultural sector to industrial sector will stop if the wage gap between the two sectors becomes zero. As we mentioned previously, the twin phenomena of labor shortages in the two industrial areas and the rise in wages match the Lewis Model. If that is so, the gap of average wages between the two sectors should increase.

In the Lewis model, the wages in the agriculture sector are lower than those in the industrial sector. The flow of labor force from the agricultural sector to industrial sector should continue until the two wage level become equal. However, the reality is that the income in rural areas is much lower than that in urban areas even though rural incomes have increased. From the Fig. 5.3, we can see that the gap between real income in rural and urban areas has decreased for a short period of time and then has increased again. The real income ratio was 2.57 when China started to reform the economic system in 1978. Then the ratio went down to 1.51 in the last years of the 1980s. However, the gap increased and it went up to 2.57 again in 2006. Since 1978, the central government has encouraged the development of township factories in rural areas and local governments have established a great amount of

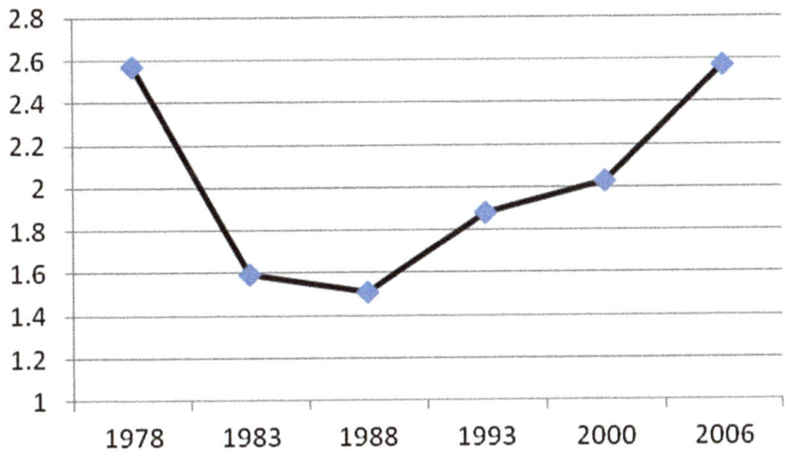

Fig. 5.3 Real income ration between rural and urban areas (Source: NBSC, China statistical yearbook, 1995–2008)

township factories. At the same time, another reform has allowed labor migration. A great amount of surplus labor went from the farm to township factories. This has contributed to the economic development in rural areas and the income gap between rural and urban area has decreased.

Figure 5.4 shows that the township factories emerged in 1978 and absorbed about 20 million people. By 1980, the amount of enterprises and their employees were kept at almost the same level. However, township factories developed rapidly from the early 1980s onwards and the amount of employees was up to over 100 million. During the same period, rural income increased considerably. As mentioned previously, productivity was very low and there was a huge amount of surplus labor in rural areas by 1978. At the same time, to develop heavy industry, central government established a lot of factories in urban regions. This made the income gap between urban and rural workers wide to 2.57. However, with the development of township enterprises, the gap narrows to 1.51 at the end of the 1980s.[18]

As we already know, China opened the coastal areas to foreign companies in 1978. The foreign investment boom emerged from the early 1990s. This can then be formatted into three main rapidly developing regions. They are the ChangJiang River, the ZhuJiang River and the BoHai Sea Ring Deltas. The total GDP of these three regions accounts for over 60 % of the entire country. At the same time, the high salary attracted migrant workers from the rural areas all over the country. However, a labor shortage only occurred in the ChangJiang and ZhuJiang Delta, but this has not occurred in other coastal areas. In this respect, this is a local problem but not common for rest of country.

[18]Minami and Makino (1999).

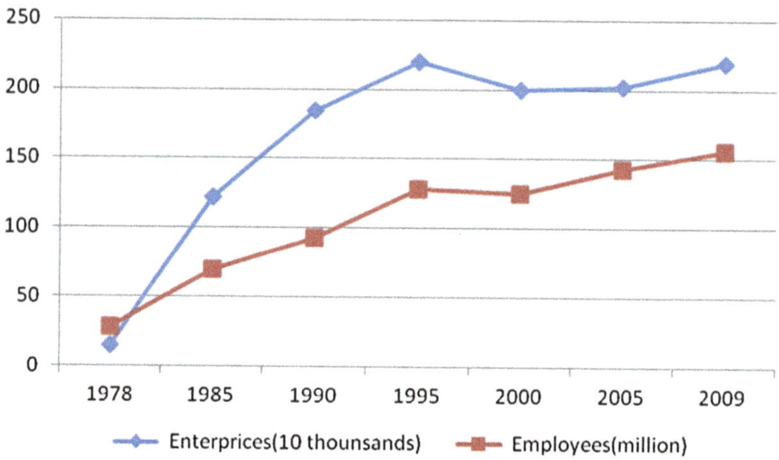

Fig. 5.4 Development of township factory in China (Source: NBSC, China statistical yearbook, 1990–2010)

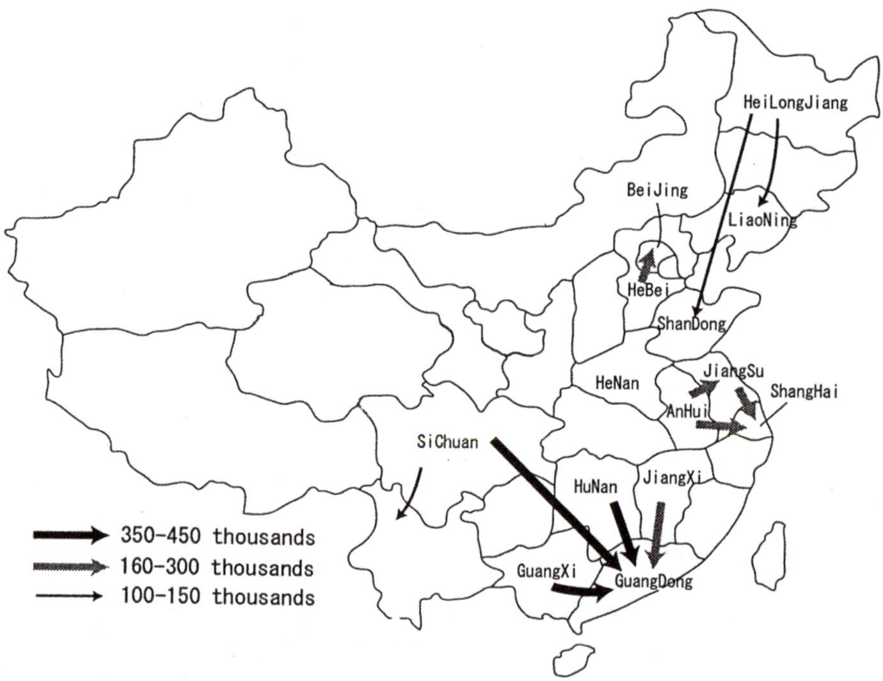

Fig. 5.5 The flow of labor migration in China (1990–1995) (Source: Minami and Makino 1997)

To analyze why this problem occurred in these two areas, we have to know what migration conditions are like in the two regions. From Fig. 5.5, we can see the direction in which labor migration has taken place and by the amount it has occurred. In the Zhujiang River Delta, migrant workers are mainly from the nearby four provinces of Guangxi, Hunan, Jiangxi and Sichuan. Each of these provinces has about 400, 000 migrants. In the ChangJiang River Delta, migrant workers are mainly from Jiangsu and Anhui provinces. Each of these provinces has about 300,000 migrants. Because the amount of migrant workers in these two areas is so big, and the local economy depends so heavily on the labor migration, the flowing back of migrant workers seems significantly affected these two regions. To prevent migration flowing back and to courage migration away to continue, local governments of these two regions have to enforce policies to protect the migrant workers.

However, in contrast, we can see China's HuKou system in which the migrant cannot live in the area of destination indefinitely. Migrants have to get a Migration Work Registration Card from their HuKou local government before their embarking a journey. At the places of destination, they have to apply for a employment license and a temporary living certificate which they should update every year. In this way, migration is not total migration but the temporary migration for the purpose of employment. In the future, they have to go back to their HuKou locality.

In the city of destination, if migrant workers want to send their children to school, they are required to pay extra money because of their temporary registration. The living conditions of migrants are poor. Many of them live in accommodation provided by their employers and it is common not to have washing facilities. Besides, migrant workers are excluded from the society insurance system. Above all, in urban areas, the migrant workers are second class citizens.

Because the migrant workers are excluded from the cities, they have no sense of belonging to the cities they are working in. Table 5.4 is the result of survey on the identity of themselves. With the survey result, we can see that among the non-urban original respondents, only 11.5 % of them think that their identity is citizen; on the contrary, 69.3 % of them think they are peasants or migrant workers. Besides, about 20 % of them are confused with their identities. This shows that only very small parts of the migrant worker have a sense of belonging to the city they are working in. However, for the majority of them, their rural registration and identity of peasant are isolating psychologically them from the city. Furthermore, as to the questionnaire of the reasons for going back to hometown, 85.2 % of the migrant worker answered that it is due to their agricultural residence. Therefore, the residence (Hukou) is one of the main reasons of their going back to hometown.

Some economists take the wage growth of migrant workers is evidence of the turning point described by Lewis Model. However, wage growth could have con-

Table 5.4 Answers of how do you think your identity in non-urban workers

	Percentage	Cumulative percentage
Peasant	30.8	30.8
Migrant worker	38.5	69.3
Citizen	11.5	80.8
Don't know	19.2	100.0

tributed to recent income growth in rural regions. Because the central government enforced the Western China Development Policy about 10 years ago and then introduced the development policy in central China, the economy in inner regions developed rapidly. Besides, the central government halted agricultural taxation in 2005 and this also helped income in the agricultural sector to increase. One result is that the income in inner rural areas has increased. Another is the migrant workers from these regions have started to go back, and this is the direct reason for why the labor shortage in coastal regions has occurred. With the survey result, we also can see that over 70 % of the migrant workers who have the plan to go back to hometown because of the increase of income in their hometown.

In recent years, central government has gradually reformed the HuKou system. From 1998, migrants with accommodation and stable employment living in at their place destination for more than 1 year were allowed the right to get citizenship there. In addition, central government has urged the local governments to improve the skills of rural labors and set up the labor market information system. However, this was just limited to small scale cities or large towns. Besides, because the enforcement of these policies was controlled by the local governments, it is questionable how much these policies have actually transformed into reality. To resolve the labor shortage, the Chinese government has also perfected the law system to protect the rights of migrants. In 2003, the State Council issued Migrant Worker Training Plans 2003–2010 which make it clear to both central and local government how training programs should be financed. In the 10th and 11th Five-Year Plans, it has been stated that the creation of fair employment conditions, social security and the protection of migrants' children's education opportunity, etc. should be provided.

These policies have contributed to the improvement of the rights of migrants. But it is not enough. One of the problems is that the actual enforcement of these policies. Anyway, the labor shortage in coastal regions has decreased competition for these local economies. To resolve this problem, local governments have had to implement migrant protection laws. Furthermore, it is not sustainable to continue under the HuKou system. With the development of the inner mainland, migrations will eventually slowdown. For coastal economies, they have developed rapidly depending on foreign investment and the availability of migrant workers. However, because of wage increases, labor intensive investment will transfer to the other regions. It is therefore necessary to transfer local economies from labor intensive to technology intensive.

5.4 Concluding Remarks

Rural China has huge population and the promotion of the use of agricultural machinery and the improvement of agricultural production technology has significantly contributed to the development of productivity in the agricultural sector. The improvement of agricultural productivity has squeezed a large number of the rural labor force out of the agricultural sector. In the Lewis two-sector model, it is assumed that the country is small and the labor force can move freely. However, because China is a vast country and it is not that simple of long-distance labor migration. The strengths of determining labor migration are in two directions. For labor migration, many factors promote it and many other factors also halt it as well.

The most important factor promoting labor migration is the economic one. In the case of the migration of rural surplus labor flowing into the city, economic factors have performed on the effect of the income gap between urban and rural areas. In addition, wealth of employment opportunities and attractive facilities such as parks; roads and public transportation, etc. have been key factors in rural labor moving into cities. Besides, convenient medical care and higher standards in medical facilities have been the main factors for which labors have considered moving into the city.

Even these factors promote the rural labor force moving into the cities, however, at the same time, many factors block the flow of migrant workers. In China, the Hukou system is one the most important issues to consider here. Although the rural labors work in cities, their Hukou locality remained in rural areas. In the city, they cannot enjoy the same social benefits as urban residents, such as medical insurance, and even their children cannot enjoy equal educational opportunities. Although they may enjoy the convenience of medical services, have to pay more than city residents because they don't have health insurance. For their children to go to school, they have to pay expensive school sponsorship fees. Moreover, their rural Hukou status reminds them as farmers, so that they cannot foster a sense of belonging to the city. Driving factors and preventing factors for migration have been competing with each other. Initially, because rural incomes are very low, the driving factors overcome the preventing factors. The promise of higher incomes and the wealth of employment opportunities have led to the occurrence of rural–urban labor migration.

In the city, migrants have some hard earned money through long times of employment. The pursuit of income has gradually faded but the troubles that work and life in the city may has cause considerable stress them, such as the issues of their education and medical expenses. In addition, with the rural economic development and employment opportunities increasing, these are some of the factors persuading migrants to return to their hometown. As a result, this has led some rural migrants eventually to return. Meanwhile, for those who are living in rural areas and have the intention of migrating to the cities, the preventing factors will gradually overcome driving factors and this is the direct cause of labor shortage.

The phenomenon of labor shortage in coastal areas is not a typical one and is not the problem for the whole country. Even migrant workers' average wages have increased and it has contributed to the increasing demand of rural areas not from the supply of labor in the agricultural sector. In China, the gap between the rural and urban region has widened but not narrowed. The development of township companies has also increased the rural income. For the future, the middle and small scaled company should be encouraged because they can ease the pressure of surplus labor in the agricultural sector and help to industrialize the country's economy.

As a historical system, the HuKou system has divided the rural and urban areas. In a historical way we can say that it has contributed to the development of heavy industry and stabilized society. But in modern China, it is interrupting economical development. As for other reforms, the government should test HuKou reform in limited regions by expanding to the entire country and gradually phase it out. To resolve the labor shortage in coastal regions, a free, fair and transparent labor market is necessary. The government should continue to protect the rights of migrants and their families. It is necessary to give the local and migrant worker the same employment opportunities.

For the enterprises in coastal areas, they have to face the labor shortage problem and considered wage increases carefully. To maximize profit and ensure the development, many coastal businesses have planning to move plants bringing great employment opportunities to labor-abundant inland areas. However, before these companies transfer to inland areas, they need to measure their markets and sort out problems in guaranteeing the supply of raw materials because a site in an inland area means they have to pay more cost in transportation if they export the products through the ports. Therefore, they have to calculate the extra cost in transportation and save on the cost of labor. In addition, the tax systems in coastal and inland areas are often another factor worth considering. Besides, the quality and efficiency of government services are also others important factors that these enterprises have to consider.

For the coastal economy, labor-intensive industries have contributed to the development of the local economy with an abundant supply of labor provided the rural areas. However, because surplus labor migration from rural areas is reduced and the level of wages should rise, as a result, coastal areas have been losing its attractiveness for investment by labor-intensive industries. Therefore, coastal areas have to transfer their economies from labor-intensive to technology-intensive industries. For China's overall economic development, developing technology-intensive industries in coastal areas and developing labor-intensive industries in inland areas and rely on abundant labor resources, this is important for China's sustained economic development.

References

China International Ecological Production Institute (2011) National Ecological Agricultural Development Research Group, on strengthening the protection of China's grassland resources. http://www.ciepi.cn/bencandy.php?fid=47&aid=717. Accessed 9 May 2011

FAO, Population and Labor Force Indicators (2004) http://www.fao.org/docrep/008/a0050e/a0050e10d.htm#TopOfPage. Accessed 20 Dec 2011

Forestry Administration of China (2005) Report of desertification in China (the third time). http://www.gov.cn/ztzl/fszs/content_650487.htm. Accessed 16 Dec 2011

Lewis WA (1954) Economic development with unlimited supplies of labor. Manchester Sch Econ Soc Stud 22(2). www.econ.yale.edu/~granis/.../lewis-manchester-2004.pdf. Accessed 12 Aug 2010

Li JN (2004) The western modern population theory. Fudan University Press, Shanghai, p 19

Meng X, Bai NS (1988) Structural change: China's rural labor force transfer. Zhejiang People Publishing House, Hangzhou

Minami R, Makino F (1999) Flowing river: migration of labor in rural China. Nihonkeizaihyoron-sya Press, Tokyo, p 10

Shenzhen Statistical Yearbook (2010) http://www.sztj.com/main/xxgk/tjsj/tjnj/201012236804.shtml. Accessed 23 Dec 2011

Solid Wastes Management Center of Henan Province (2011) http://www.hnep.gov.cn/gfgl/tabid/683/InfoID/20376/Default.aspx. Accessed 16 Dec 2011

Xinhua Net (2011) State council on improving the project of grain for green. http://news.xinhuanet.com/newscenter/2007-09/10/content_6698742.htm. Accessed 8 Oct 2011

Chapter 6
Sustainable Social Security System in Rural China

6.1 The Ageing Problem in Rural China

The twenty-first century has an ageing population and the problem of ageing is a global phenomenon not simply confined to one area or state. In 2000, out of 204 countries and regions in the world, 72 countries have shown they have ageing population; the proportion of the population is up to 43.36 % of the entire world's population.[1] All the developed countries are facing the problem of an ageing population and many developing countries are entering or will soon enter the situation of an ageing society. Internal Revenue Service defines ageing populations as people over the age of 65 years.[2] Normally, the share of population over 60 years accounts for more than 10 % of the total population, or population over 65 years accounts for more than 7 %, these countries or regions are called as "old age societies". In this chapter, I will use the common concept of ageing which concerns the proportion of the population over the age of 65.

China entered the "ageing society" in 2000. According to China's fifth population census in 2000, the ageing rate of this country is up to 7.1 %, this is the indication that China is entering the "ageing society". Moreover, China is the world's most populated country and has the most ageing population. The ageing population in China accounts for one-fifth of the total world. So researching China's ageing problem is extremely important not only just for China, but also for the entire world.

One of the important features of China's ageing problems is the large difference between regions. Some regions such as Shanghai entered the ageing society early (Shanghai was in 1979). Since the third population census in 1982, the development of ageing has shown an imbalance in characteristics between urban and rural areas.

---

[1] Ibid; p. 24, see note 25, Chap. 1.

[2] Ibid; p. 23.

B. Qin, *Sustainable Development in Rural China*,
DOI 10.1007/978-3-662-46476-2_6

Table 6.1 The ageing degree in three population census (%)

	Average	City	Town	Village
The third time (1982)	4.91	4.68	4.20	5.00
The fourth time (1990)	5.57	5.38	4.42	5.74
The fifth Time (2000)	7.10	6.67	5.59	7.50

Source: NBSC, population census of 1982, 1990, 2000

From Table 6.1 we can see that in 1982, the ageing rate was 5.0 and this was higher than in towns or cities, therefore rural had the highest degree of ageing populations. By 2000, the ageing rate in cities and towns was up from 4.68 % and 4.20 % to 6.67 % and 5.59 % respectively. On the contrary, the ageing rates in rural areas rose to 7.5 % and were still the deepest ageing region. As for growth rate, rural areas were up to 2.5 % in this period, far more than 1.99 % and 1.39 % in towns and cities. This shows that rural regions not only have the deepest ageing problem but also the ageing process speeded up the fastest.

Even there are lots of reasons for the difference in ageing between urban and rural populations; the main reason is due to inter-regional labor migration. From the analysis in the previous section, due to government released control on labor migration, a large number of rural labors rushed from the agricultural sector into the industrial sector in urban areas for higher wages and abundant employment opportunities. The main part of the workforce consisted of young adults but elderly people who were left in the villages. In 2000, China Statistics showed that in more than 100 million migrants, 73 % are from rural areas. And in the floating population, the average age of migrants was 33.6 years old and 94.43 % of these people were aged less than 60.[3] The share of the property of the elderly population in rural areas increased and this is the direct cause for the ageing rate of rural areas to be higher than that of the urban ones.[4]

In addition, the ageing situation is a major for the labor-outflow from the provinces and it can also reflect the impact of the labor movement. As I analyzed in the previous chapter, since the late 1980s, China has opened its economy to the rest of the world and has kept up its rapid economic development, especially in coastal areas. As the leading industry was mainly "Three-plus-one",[5] these labor-intensive industries, generated a lot of demand for labor. For higher income and work opportunities, a large number of inland labor migrated to the coast areas, Liaoning, Anhui, Hunan, Guangxi and Sichuan are the main provinces for labor

[3]Ibid; pp. 51–52.

[4]The data of Sichuan Province in 1982 include Sichuan Province and Chongqing city. Chongqing became a municipal city in Sichuan province in 1997. Because the ageing factor of Chongqing in 2000 was 8.01, the ageing growth rate of Sichuan Province should be slightly higher than in the table considered for Chongqing.

[5]A form of corporate trade Mainland China tried to create an atmosphere of openness at the beginning of the expanding economic period. This included the customary manufacturing of materials, designs or samples supplied and compensation trade.

Table 6.2 Ageing degree of major labor outflow regions (%)

Region	1982	2000
National average	4.91	7.10
Liaoning province	4.81	7.88
Anhui province	4.08	7.59
Hunan province	4.97	7.47
Guangxi province	5.11	7.30
Sichuan province	4.68	7.56

Source: NBSC, population census of 1982 and 2000

outflow. By observing the ageing conditions in these provinces of mainly labor force supply with Table 6.2, we can see clearly that the labor outflow has influenced the ageing process.

In 1982, China had just opened up its economy however the development has not yet entered the stage of rapid development which can be seen in coastal areas, and therefore the local labor supply could meet local demand. In addition, the government has not released controls on labor migration, and labor migration between regions is not obvious. In these provinces, except that of Hunan and Guangxi which was slightly higher than the national average, the ageing rates of other provinces were all lower than the national average.

After that, the economy in coastal areas developed rapidly and the income gap between coastal and inland areas became wide. Young worker flew from inland regions especially those provinces after said to coastal areas and this contributed to these region's age process. In 2000, the fifth population census data showed that the ageing rates of these provinces became higher than the national average. Furthermore, the growth rates in the ageing populations of these provinces were much faster than the national average. Especially Liaoning Province which increased from 4.81 % to 7.88 %, Anhui Province rose 3.51 % from 4.08 % to 7.59 %, compared with the national average of 2.19 %. Certainly, most migrants in coastal areas were from rural regions.

Compared to cities, the ageing problems put greater pressure on rural areas. This is because a large number of young adults flow into the city to work, the decline in support for older people in rural areas makes family care more difficult. Meanwhile, as most rural areas have not established an old age pension system, peasants lack the necessary social security they would get from having a pension. In 2011, the central government proposed to establish an old-age security system in rural areas in the Twelfth Five-Year Plan of China's Economic and Social Development and it is the view that solving the problem of rural old-age security as an important issue. Therefore, the study of ageing issue is very important for the sustainable development in rural areas. However, for ageing rural China, establishing the social security system is extremely necessary. Because old-age pension security and medical insurance system are the most important contents of the social security system, I will analyze these problems in the second and third chapter.

6.2 Research on Pension System: Sino-Japan Comparative Study

Because almost all developed countries are entering the ageing society, to address social problems caused by the ageing issue it is one of the important topics for these governments. In the researching of China's ageing issue, the comparative method of studying one of developed countries should be effective. This is because, as previously analyzed, even though there are many reasons for a country's ageing population, the economic factors especially the industrial developments of the state should have a very close relationship with ageing problems.

In addition, developed countries entered into the ageing society much earlier, their policies should be a reference point for China to address ageing issues. In 2000, 17 % is the highest ageing rate in the world and the countries that have reached this point are Sweden, Italy and Japan. Japan entered the ageing society in 1970 and it was of precisely the most mature period of rapid economic development. This is very similar to China. Furthermore, in the agricultural sector, as is the case for China, Japan also reflects the small family-run model. This has important implications for the study of the sustainable development in rural China. For the above reasons, I have selected the comparative research methods for this section. Firstly, I will compare the reasons for the ageing population and the problems it brings from the aspects of changing in the industrial structure, labor migration and decrease in the birth rate. Then I will analyze the pension systems in both countries and explore the sustainable path for ageing rural China.

6.2.1 *The Analysis of Ageing Problems in China and Japan*

Firstly, let us look at the change in the industrial structure in China. The first industrial share of the GDP appeared to show a brief rise from 1978 since China had implemented a contract responsibility system in rural areas. After then the potential share of agricultural production has been narrowing subsequently due to the diminution in marginal productivity. The share of agriculture in GDP has gradually declined and reached 11.31 % in 2008. Compared to China, Japan has shown a similar trend. In the years since the end of the Second World War, Japan' industrial structure has changed considerably. The first sector's share of GDP took a downturn while the second and the third showed the opposite trend. The first sector went down from 21.9 % in 1955 and reduced to about 11 % 10 years later the same as the 2008 level for China because of industrial development. Then it gradually reduced to 2 %. As for the second sector, it increased from about 30 % in 1955 to over 40 % in 1970. Then it decreased slightly and stabilized at the 35 % level. On the other hand, the third sector went upwards from 40 % to over 60 %.[6]

With the change in the industrial structure, the employment structure changed too. Based on the style mass-production, productivity in the industrial sector is dramatically higher than the first sector and its contribution to economic development is greater. Industrial development generated a great demand for labor. In Japan, the popularity of agricultural mechanization and the improvement of agricultural production technology produced a large amount of surplus in rural labor. On the contrary, a large number of hidden rural surplus labors were to be released in China mainly due to changes in the agricultural production system. Whether in Japan or China, this release in the labor force from the agriculture to the industrial sector met the demand in the labor supply and contributed to industrial development.

The era in which the labor migration process took place in Japan was about 40 years earlier than that in China. As analyzed in the previous section, labor migration is often more common in regions where there are high-level incomes and rich employment opportunities. In Japan, the most economically developed areas are concentrated in Tokyo, Nagoya and Osaka, the net-migration to these three zones appears at its highest populated areas. Especially in 1963, over 600,000 people migrated into densely populated cities.[7] In China, it has been concentrated in the three regions of the Pearl River Delta, the Yangtze River Delta, and the Bohai ring economic zones. From the 1990s, many foreign factories emerged in the coastal areas and increased the huge demand in the amount of labor. The migration from the inner mainland to the coastal region emerged. From Fig. 5.2 in previous chapter, we can see that migrant workers amounted to 61.99 in 1993. After that, migration continued and the amount of migrants rose up to 136.49 in 1997.[8] Whether in China or in Japan,[9] labor migration from the rural to the urban areas became one the most important reasons for the ageing populations increased in rural areas.[10]

Meanwhile, another very important reason for ageing is the natural reduction in the population. This means that the birth rate is less than the mortality rate. Figure 6.1 shows the trends in both the birth and mortality rates in Japan. From it, we can see that from the 1970s, the birth rate decreased continually and the mortality rate increased at the same time. Finally in 2007, the mortality rate surpassed the birth rate and the population shrunk.

There are lots of reasons for a low birth rate. Firstly, in urban Japan, many families choose not to have many children due to the high cost of feeding a family and high education costs. Because there was no regulation on population migration, migrants from rural areas automatically became citizens. However, living in cities

[6]Qin (2009).

[7]Qin (2010).

[8]Ibid; see note 3, Chap. 5.

[9]Statistics Bureau of Japan, Summary Table of Vital Statistics 1890–2004, http://www.stat.go.jp/data/chouki/02.htm, accessed on 2009.10.18.

[10]Ministry of Health, Labor and Welfare of Japan (MHLW) (2009).

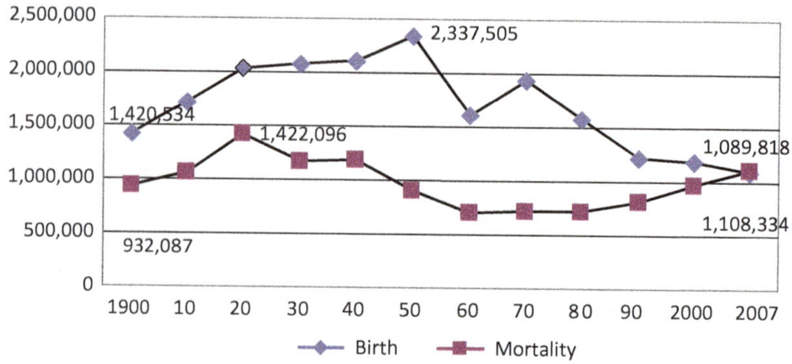

Fig. 6.1 Birth and mortality population in Japan (Source: Statistics Bureau of Japan, 1890–2004 and MHLW 2009)

means poor housing conditions and high living costs. Extra room for bring up children, the expense of feeding children and educating require citizens to spend a lot of money. This caused the number of low income families to rise.

At the same time, Japan became a highly-educated society as the industries changed from labor-incentive to knowledge-incentive industries. For their children's employment in the future, Japanese families have to pay extra money to send children to college. So the burden incurred by the cost of education for the families is becoming heavy. The educational expenditure on education for households was just over 2 % in 1965 and then it went up by 2.5 times of that in 2004 to 6 %.[11] Leibenstein[12] believes that there are the huge costs incurred from child-care including the direct expenditure on each child and the opportunity cost of time spent actually on raising children. In addition, a couple's fertility rate has fallen regardless of changes to family income.[13] There is an increasing burden placed on urban households to choose whether or not to have fewer children or no children at all.

Secondly, another reason for a low birth rate is rising of the age at which people get marries. Due to a highly educated society, more young people choose to go to college and the age at which they tend to marry has risen. If people go to college after they have graduated from senior high school, the age at which they will marry will be raised by 4 years. In addition, after graduation from university, most people will work for many years and be prepared to enjoy life more beforethey get engaged

[11]Ibid; see note 4.

[12]Leibenstein (1975).

[13]Liu (2004).

and so the average age for marriage tends to be higher. Japanese men used to marry at about 26 and woman at 23.5 in 1950 but the age rose to 30.5 for men and 28 for women in 2005.[14]

Furthermore, with the Industrial change in Japan from labor-intensive industries to technological-intensive ones and the development of the third sector, it is possible for women to be more active in the economic development of the society. They received salary and so no longer have the need to get financial support from their husbands. Some women choose to lead a single life, even though one day they may marry. So the age at which they will marry is getting later. Therefore, the age at which they will have children is getting older as well. The mean age of mothers giving birth for the first time was 24.4 in 1950 and increased to 29.1 in 2005.[15] Late marriage and getting birth at an older age has contributed to the decrease in the birth rate.

6.2.2 *Reform the Social Security System*

In Japan, both government and the media consider that the country will be confronted with a lot of problems caused by the ageing population and low birth rate in the future. One of the main problems is that the social security system would collapse. The social security system is designed to maintain the citizens' income after they have retired. Under the pension system, young people pay money to the government when they are at the working age, and get the pension from the government after they have retired. In most countries in the world, the so-called "pay-as-you-go" system has been adopted. In Japan, the same social security was also adopted.

Actually, the government pays the working generation's money to the elderly generation. After the current working generation has aged, the retirement pension would be from the next working generation. This is because the government expects the population to continually increase. So far up until now, it has never been problematic because the population has also been increasing. However, once the population decreases, the current system cannot be sustained. Figure 6.2 shows the income and expenditure balance of pensions in Japan. We can see that from 1961, pension income was greater than expenditure. The balance went upwards and peaked in 1995. After that, it went down and went into negative from 2002. The current social security system was designed with continuingly expanding population in mind but this isno longer the case.[16]

---

[14]Ibid; see note 6, Chap. 5.

[15]Ibid; see note 6, Chap. 5.

[16]MHLW (2012).

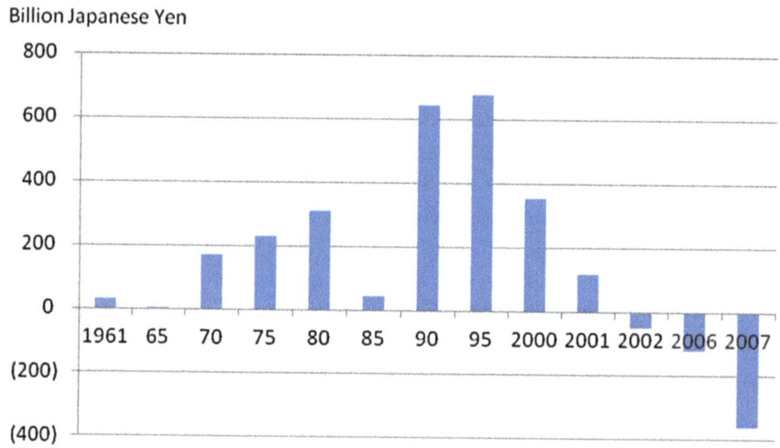

Fig. 6.2 The pension balance in Japan (Source: MHLW, report on pension fund management, 2010)

In EU countries, governments are facing a similar problem. In Italy, spending on the "pay-as-you-go" system as a share of gross domestic product is expected to peak at 15.8% in 2032. In Germany, forecasts indicate a similar peak, although these projections do not take into account that this year and the next, pensions will be increased, though not in line with average net earnings, only by the rate of inflation. In France, where population decline will be less pronounced, state-funded pensions are expected to rise to 16 % of GDP in 2040.

To combat this looming crisis, many governments have been trying to look for new sources of finance. One approach is to widen the number of younger people in the population generation and allow greater immigration. Young skilled immigrants can expand the workforce by raising productivity. Another possible remedy is by generating taxation. But the trouble here is that tax financing weakens the links between contributions and benefits in the pension system. The most fundamental solution is to reform the existing pension system.

In Japan, the pension system covered the whole country and rural areas have not shown that ageing has become a social problem. On the contrary, as pension system has not been established in most parts of rural China, it has not been a very prominent in issue. However, according to the proverb "it is easy to map in white papers", it will become an important issue in rural China in the future and China will have to learn the lesson of Japan.

6.2.3 The Sustainable Pension System in Rural China

In rural China, family has always been considered the main source of support for the elderly as the pension system's coverage rate is so low and traditional family culture has always been so deeply rooted. However, in recent years, family support for the aged has been weakening continuously. There are many reasons for this. Firstly, as shown in the previous analysis, the situation of young adults migrating to urban areas has meant that most elderly people have stayed in rural areas. The degree of ageing in rural areas has become higher than that of the urban areas. At the same time, this has caused a weakening in family support for the elderly. Secondly, the implementation of the one-child-policy has not only caused a rapid decline in population, but has also changed the family structure. At present, Chinese family commonly consists of four elderly people, one young couple and a child known as the 4-2-1 composition of inter-generational structure. This inter-generational structure shows that it is completely impossible for a family to give support to aged family members. Again, due to the current trend, the young and elderly tend to live separately and the increase of the nuclear family has meant that the self-maintenance of the elderly has gradually increased.

As a supplementary measure to the family planning policy, the government should provide auxiliary assistance to FSFA. For a young couple, it is very difficult in not only in an economic sense but also in terms of energy to bring a child whist supporting four elderly people. In reality, some local governments have subsidized families based on the award of family planning points. For example, the Guiyang government has issued such subsidies in 2002.[17] Based on this fact, once men have reached over 60 and women over 55, single daughter households can receive the pension of 60 RMB and one-child and two-daughter sterilized households can receive 50 RMB every month. The government of Nanjing government has also implemented a similar provision in 2003.[18] This local government has declared that rural residents of single daughter households can receive a pension of 25 % of the Minimum living standard once they have reached over 60 years old. For those households in which a child's premature death has occurred and which have not had the birth of another child or have adopted any more children, they will get 50 % of the rural minimum living standard every month.

Effectively, such measures can help ease pressure on the family which regards support for the elderly and should serve as a unified policy across the country. Again, the central government should increase the amount of subsidy as far as possible to ease the family burden concerning support for elderly relations. However, the family should not be the main source of support for the elderly but, it should be combined with other measures.

[17]Guiyang Local Government, 2002, "The Ligation Method of Providing a Pension to One-child and Two-daughter Sterilized Families in Rural Guiyang".

[18]Nanjing Local Government (2003).

Arable land is considered another source of income for aged rural residents. Because rural residents are contracted by local governments and have the right to use arable land, some scholars believe that the arable land also has the function of providing support for the elderly in rural China. Actually, since the implementation of the household responsibility system in 1978, the land has become the economic backbone of rural households. A source of income, job security, and pension security are all maintained on the land to which the household has been contracted. However, the protection level is low and the support capacity is limited. In addition, because the land contracted by farmers is always under the threat of expropriation as the process of industrialization and urbanization accelerates, the old-age support functions are weakened.[19]

Because there is an absence of a social security system in rural China, the land output is the key point for examining whether the land can assume the function of old aged support or not. Agricultural income in a rural household accounts for about 45 % of the whole income, and this is only the equivalent of minimum standard of living.[20] However, this income level is considered without taking labor input into account. If we exclude the labor costs, the land output will be about zero or negative. This shows that the land has only played a significant role in the case of self-employed for farmers and the result of land management is zero or even deficit.

More profound is that the previous analysis is only a general analysis without considering the problem of labor's age. In fact, the ability of the elderly to cultivate land in rural China is very limited. Because they just play a supporting role in agricultural production, land management is very difficult or even impossible for them without a young labor force in the family. Therefore, it is simply negligible that the land management should support the old aged in rural China. If agricultural land were to be expropriated, the situation would be different. As compensation for land expropriation is different in various regions, the ability of land to provide old-age support is also different. As the analysis has shown in the previous chapter, when the land was converted from agricultural to other uses, compensation went generally towards currency payment and the provision of social security. The purpose of this compensation will be for land, resettlement subsidy, crops compensation and ground attachments compensation if the compensation of acquired land is in accordance with its original purposes.

Because the sum of compensations payments should not exceed thirty times the average annual output of the agricultural use and the money to the small-holders is always very little, this very limited compensation cannot meet the long-term survival for landless peasants, not to mention the support that they will be in old age. For the peasants that have lost their land, being absorbed into the urban social security system is the only sustainable way for them to survive as they reach old age.

Support for their economic life and pension terms should be sustainable. Since land takes on the function of a production tool and also the old-age security, peasants

[19]Xiao (2007).

[20]Han (2010).

who have lost their land have to get expropriation compensation. On one hand, they should be paid in cash to maintain the source of income for their economic survival. On the other hand, they should to be provided with some old-age security services. Only be receiving compensation and pension protection, can their economic survival and future pension security is be guaranteed.

The establishment of a rural minimum living security system has been considered another supplementary measure in which to solve the old-age pension problem in rural China. This policy is an integral part of the current social security system in almost all developed countries. For the purpose of protecting the basic rights of nationals, it is essential to maintain the national's minimum living standard and as the last defense towards building a proper social security system. When the national income level is below the minimum living standard, it cannot maintain adequate living standards and so assistance should then be provided by the government.

In China, Minimum living security for urban residents has sieved the attention of the government and has subsequently been implemented. After the success of the Shanghai pilot scheme in 1993, most economically developed coastal regions gradually built up this system. Since then, the central government has issued the State Council notice on establishing the Minimum living security system for urban residents[21] in 1997 and also issued the minimum living guarantee files for urban residents[22] in 1999 to widen the policy at a nationwide level. At The end of 2004, more than 20 million rural residents enjoyed this guarantee and the policy realized its basic security function.

However, the rural minimum living security system is far from resolved. Although China's economic reform was primarily focused on the movement from rural to urban, economic distribution went in the opposite direction. On the one hand, as an important part of the security system, the time needed to implement the minimum living security system in the rural areas was later than in the urban ones. In fact, 10 years later than in urban areas, so the central government issued the State Council notice on establishing the Minimum living security system in urban areas in 2007.[23] On the other hand, the scope of its protection was limited within some areas. Before its promulgation, some provincial governments piloted the system and gradually established their own local security systems. However, it was not as popular as those established in the cities.

One important reason is that the number of poor in the rural population is larger and tends to be concentrated in those rural areas. China's rural poor population was about 40 million in 2009. Again, as the funding of the Minimum Living Security System comes from local finances, this put pressure on local governments especially in the economically underdeveloped western regions.[24] An easily overlooked

[21]Chinese Central Government File [1997], No. 29.

[22]The State Council of China, Decree No. 271.

[23]Chinese Central Government File [2007], No. 19.

[24]Article 5 of notice: The funding of rural minimum living security should be from the local government mainly.

problem is that of rural poverty standard. The World Bank's poverty line is based on a daily consumption of U.S. $ 1.25 per capita. However, China's current poverty line for rural residents was that net income of 1,196 RMB annually per capita and this was determined in 2008. It works out as only $ 0.89 per day at this level of purchasing power based on parity calculations. It is far lower than the standard set by the World Bank.

Considering the economic factors such as rising prices, the Chinese government will increase the level considered to be the poverty line to 1,500 RMB of net income annually in its "Twelfth Five-Year Plan". Effectively, the number of poverty in the population will also increase substantially especially in the economically underdeveloped western regions. For instance, Guizhou ranks second by having about 5 million poor people in its populations in 2010, this will rise to 10 million after the introduction of new standards and thus the number would have doubled. The new standard will then make the unresolved issue of popularizing minimum living security system more difficult.

Some scholars believe that a more feasible approach is to invest in poverty alleviation funds which will be directly paid into the minimum living security system. Therefore the sum for the poverty alleviation fund from central and local governments has reached about 30 billion annually. If the funds are distributed directly to the hands of the poor, should amount to about 750 RMB per capita. Add the average annual income of 500 RMB, the income level will be covered and the minimum living security problem should then be resolved.[25]

Actually, the method of poverty alleviation in China is guidance and it has helped a large number of poor out of poverty. However, once the fund has directly distributed to the rural poor, there is a high likelihood that residents just out of the poverty will return to being poor and that the result is a significant increase in poverty. At that time, the poverty alleviation fund will be in seriously lacking and the form of direct payments will experience paralysis. Therefore, China should not only continually alleviate poverty by reducing the number of poor in the population but there should be an attempt also to popularize the minimum living security system in rural areas. Especially for the economically underdeveloped regions, the central government should increase financial support, strengthen local finance and take major responsibility.

In summary, relying solely on land revenue support for the family cannot solve the issue of rural old-age pensions. Meanwhile, the minimum living income security system can only serve as a supplementary measure. Therefore, the Chinese governments at both local and national levels have to build a broad social old-age security system in rural regions.

---

[25] Wu and Zhou (2004).

6.2.4 Pension Fund Choice for Rural China

For the establishment of a rural old-age security system in rural China, the most critical issue is the collection of pension insurance funds and the methods used to do so. The basic principle is to maintain the income and expenditure balance dividing these in terms of lateral and longitudinal balance. Lateral balance means consistency between the fund and the sum of the pay in the current year and longitudinal balance means the consistency between the payment and the pension in the whole pay period of a participator. Based on these principles, the fund mode can be divided into a "pay-as-you-go", a "fully funded" and a "partially funded" model.

"Pay-as-you-go" is where the funds are managed based on the principle of horizontal fiscal balancing. In this system, the total pension expense of the current year is initially estimated and then it is apportioned into the sum of a payroll, finally determine the premium rate is determined. This pattern reflects the maintenance of intergenerational relations. One of the advantages of this model is that it is convenient to operate and many developed countries have built their pension systems on this very model. It is fully feasible in an era of population increase. However, this system cannot be sustained once an ageing stage in the population has been entered.

As previously analyzed, under the "pay-as-you-go" system, Japan's pension income has been steadily increasing and has reached a peak of 6,790 million JPY in 1995. Then it has been in decline because of the population has been decreasing and finally it has appeared to have shown a deficit of 485 million JPY in 2002. Since then, the country's pension fund has been in deficit and this is one of the most troublesome problems for each successive ruling government. Just like Japan, other developed governments such as the European Union countries have also faced the same problem.

A fully funded mode is where funds are managed based on the principle of a vertical fiscal balance. In this system, the participator's health condition is forecast and the economic development indicators can be formulated within an entire life period. Then the amount of pension payments are measured out and spread over his (or her) entire insurance period. After participators have retired, they can then get a pension from their personal pension fund accounts. This model can avoid the ageing problems caused by inter-generational relations and there is no necessarily to adjust the premium rate because it is pre-measured and stable.

However, the shortcomings of this model are obvious. Firstly, it is weak of freemasonry characteristics because it reflects a self-funded and does not work a social redistribution function. Secondly, it is quite difficult to predict the amount of pension as not only it does involve many predictors but it also involves considering the span of time that people's whole life spans cover. Again, the Fund is vulnerable to government intervention, such as being diverted to other purposes or corruption.

A partially funded system is an integrated form of "pay-as-you-go" and "fully funded" system. It is competence with the principles of a horizontal and longitudinal balance in payments and based on the principle of keeping a balance between income and expenditure at various stages. Under the premise of leaving certain reserve funds, the pension rate depend on the spending needs of a certain period.[26] The amount of Reserve Fund is variable. It is accumulated in the period before the population has aged and it is consumed in the period after the ageing peak has been reached. One of the advantages of the partially funded mode is that it has flexibility. On one hand, it can avoid great risks compared to the fully funded mode. On other hand; it also alleviates the problems of a lack of reserve and uneven burden compared to the "pay-as-you-go" model. However, this system is difficult to implement, especially in the mastery over varies ratios. If the standards and ratios are set incorrectly, instead of getting the desired effect, it will lead to a substantial increase in management costs. Therefore, it needs great caution in the designation of the system especially in as far as a quantitative analysis goes.

Because the level of economic development and the family income levels in rural areas are very low, relying on the land solely as a source of revenue cannot solve the problem of the rural old-age pension. At the same time, due to population decline and changes in the structure of the family, and especially the migration of young labor enter into the cities continually weakens the ability of the family to provide security for residents in old age. In addition, a rural minimum living security system can just serve as a supplementary old-age security measure. Therefore, to solve the problem of an old-age security system in rural China, it is necessary to create multiple, multi-level security systems. Firstly, family support should be strengthened. Secondly, the government should protect the land contracted to peasant families and establish a land requisition compensation policy to improve the level of compensation from expropriation. Again, governments have to establish a rural minimum living security system to protect the primary right to life of the rural elderly. On this basis, it is necessary to establish and popularize social old-age security systems in rural China.

Under this premise, the China governments have to establish a sustainable pension fund. As the precious analysis has shown, because Japan adopted the "pay-as-you-go" system, the pension funds went into deficit due to an ageing population, therefore we can say that this mode is not suitable for ageing rural China. On the contrary, because peasants are self-employed, fully funded or partially funded models are the more feasible options. In addition, the Chinese government should establish a rural pension insurance reserve fund earmarked specifically to solve the problems of an ageing rural population. The amount of reserve funds should be adjusted based on the development of the ageing process in rural areas. Distribution should be flexible based on the degree of ageing in each province and the fiscal capacity of local governments.

[26]It is often 5–10 years.

6.3 The Medical Insurance Health System in Rural China

6.3.1 The Situation of Healthcare in Rural China

The medical insurance system is another very important issue for sustainable development in rural China. Since the low level of economic development in rural areas, health care costs are heavy burden on small-holders. According to the Ministry of Health Statistics of 1999, the net income of peasants has increased by 2.2 times over 10 years. On the contrast, the average cost of outpatient and hospital charges increased by 6.2 times and 5.1 times over the same time span.[27] As rural residents cannot afford medical expenses, about one third of patients in rural areas should be hospitalized receive no hospitalization.[28] In rural areas, the condition of "suffer slight illnesses, delay serious illness, life-threatening illness to the hospital" is common. Poverty due to illness is very serious, especially in western China, these peasants number reached to 300–500 million and 70 % of poor households in rural areas are as a result of illness.[29]

However, before economic reform in China, the health care system in rural areas has became very popular. Since the founding of the country, the Chinese government has been committed to the establishment of rural health care and has begun cooperative medical care system started in 1955. After the implementation of the rural people's communes in 1958, the national cooperative medical coverage reached 10 %. Since then, the central government has increased its efforts to popularize the system across the country and coverage was up to 46 % in 1962.

However, since 1968 there have been happened a number of natural disasters over a period of 3 years and so the coverage rate was reduced, by 1976, it was resumed and up went to 90 %. The medical insurance system in rural China achieved a very high degree of popularity. At that time, relying on the rural commune, China formed a perfect rural health care system consisting of the county hospital, commune health centers, and village clinics. In China, the situation of "there are medicine and doctors where people are living" and "minor illnesses within the village, serious illness within township" was the norm. The Chinese people's health care system had been greatly enhanced. China won wide praise for the most comprehensive health care system in the world at that time. The World Health Organization gave a high evaluation of China's extraordinary achievements in rural health care and actively recommended the experience to other developing countries.[30]

Figure 6.3 shows the development of the rural medical care system in China. In 1978, China implemented a series of rural economic reforms and adopted the "household responsibility system". From then on, the family became the basic

[27]Wang (2001).

[28]Ministry of Health Statistical Information Center, 2004.

[29]Zhang (2002).

[30]Li (2009).

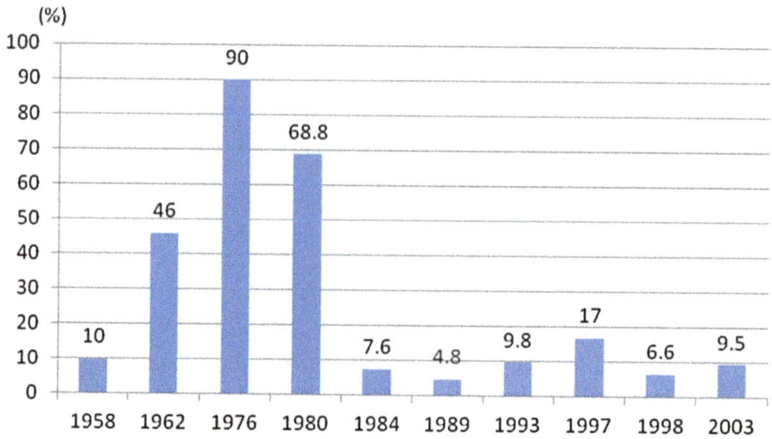

Fig. 6.3 The share of CMS village in China (Source: Li 2009)

production unit in rural areas. Since the disintegration of the collective economy, rural cooperative medical care system lost its backing and collapsed rapidly in most rural areas. By 1984, the rural cooperative medical care coverage decreased drastically to 7.6 %. After the 1990's, the government re-recognized the importance of a rural medical security system and released[31] in 1993. In this file, the central government proposed the development of cooperative medical system, but did not take any action on it. By the end of 1997, the cooperative medical system just covered only 17 % of administrative villages.

6.3.2 *The Equity Issues Between Urban and Rural Areas*

As analyzed in the previous chapter, as there is a large economic gap between urban and rural China, there exist two sector characteristics in all social systems. There is also a huge urban-rural gap in a medical insurance system. Firstly, the allocation of health care resources between urban and rural is unequal. Medical staffs with higher technology are crowded into the city's large hospitals. In contrast, village clinics or individual doctors are the main resources that rural residents use. Most of the village health workers have not participated in formal training, and a considerable part of the village clinics are without the necessary disinfection equipment. Secondly, compared to cities, the share that the central government invests in rural health is low.

[31]Chinese Central Government, 1993, "On Several Decisions of Establishment of a Socialist Market Economic System".

From 1979, the government generally abandoned the responsibility of health care for farmers. Even though the rural area represents 80 % of the entire population, medical expenses for the rural areas only accounts for 16 % of the total government health budget. On the contrary, 20 % of whole population in cities takes up 84 % of the medical budget. Thirdly, health care expenditure differs greatly between urban and rural areas. Because rural residents mainly shoulder the personal burden of medical expenses themselves, the urban health insurance system has improved gradually. The average medical cost of urban residents per capita was 2.6 times that of average rural residents in 1995. However, after 3 years in 2001, the gap widened to 3.6 times.[32] This means that the urban-rural gap performance does not only reflect in residents' income, but also a show up in health care services and it is right to say that this gap is continuing to widen.

The lack of Medical insurance system in rural China harms the peasants' health seriously. As previously analyzed, as the young labor force transfers to the cities, the degree of ageing in rural areas exceeds that in urban areas and becomes even more serious. The medical needs of the increasingly elderly population cannot be met. As a result, rural health has deteriorated. The 2-week incidence rate increased from 128.2 % in 1993 to 137.1 % in 1998 and rose to 139 % in 2003.[33] Because the rural poor cannot afford medical expenses, about one third of patients in rural areas should be hospitalized but receive no hospitalization. In rural areas, disease is one the major factors of poverty and 70 % of poor households are a result of illness.[34]

To address the rural medical security issue, the central government enacted.[35] In this file, the central government proposed to establish a New-type of Rural Cooperative Medical System (NRCS) in a nationwide rural area scheme and planned that the new rural cooperative system covered all of the rural residents basically by 2010. Differing from the mandatory of the old system before 1978, one of the principles of NRCS is voluntary participation. The other difference is that the central government does not require all the local governments to establish NRCS at the same time, but provides them with pilot lessons and improves the system steadily in accordance with real life situation for their areas.

Solving the funding problem is the most important thing for the Establishment of a sound social security system. Compared to Urban health insurance funds from companies and individual wages, the local residents are self-employed and their income is low. So the funding source of NRCS is lacking. To solve this problem, the central government clearly states the implement combined funding mechanisms and the source is from individual peasants who pay with government support. Firstly, every insured rural resident pays 10 RMB annually. Secondly, the central government will spend NRCS 10 RMB in subsidy per capita every year from 2003

[32] Ministry of Health of China (2004).

[33] This is a view point of health service needs from the residents and the incidence includes the prevalence determination by medical staff and the respondents' own feelings of "discomfort".

[34] Ibid; see note 29.

[35] The Chinese Central Governmental File (2011).

to peasants in central and western provinces who participate in the scheme. Finally, every provincial government has to subsidize the participator of NRCS, they can even determine the specific local subsidy standard, but the allowance should not be less than 10 RMB per capita annually.

By the end of 2010, the number of NRCS participator was up to 835 million and the participation rate reached to 95 %.[36] But a lot of problems emerged. First, because of the moral crisis of health care institutions, the participators of NRCS have to pay more medical expenses and this has caused a heavy burden on them. After the rural hospitals obtained the certification necessary to have fixed medical units, the hospitalization and outpatient fees generally rose. Such a phenomenon has become common place such as the serious treatment of slight ailments, disease-free medication, and irrational drug use. This phenomenon increased the burden on NRCS participators. Especially, some phenomena appeared whereby the NRCS participators bought drugs in the designated hospital; the fees they need to pay for these are still more expensive than purchasing medicines at a drugstore even though reimbursement is available.

Secondly, the government failure has caused high administrative costs and inefficiency. Each Local government lacks the management to operate NRCS and these institutions are in disorder and ineffective. In some areas, NRCS is managed by several institutions but the responsibility for its operation is not clear. As the management agency's operating costs are too high, the funds should be spent on medical costs instead. Again, reimbursement involves a lot of requirements and complicated procedures. Many rural residents have to travel to several departments for the reimbursement. Furthermore, the differences between rural and urban areas that have already been mentioned have not been eliminated.

6.3.3 Field Survey of Zhanjiang's Integration Medical System

To resolve the above problems, the central government encourages localities to explore and establish the supplementary medical insurance scheme under which of improvement are to be made to NRCS. One of efficient methods is Joint qualified commercial insurance agencies to launch a new rural cooperative service. In January 2009, the Zhanjiang City government worked with the Chinese People's Insurance Company (CPIC) to jointly establish a rural-urban Unified Health Insurance System (UHIS). After more than a year in operation, UHIS achieved good result: "none-increasing of government investment, non-rising of resident's burden, improving level of medical care, broadening coverage of insuring". For the study of sustainable social security system in rural China, I interviewed theGeneral Manager of the CPIC

[36]Rong (2011).

Zhanjiang Branch in May 2011 to research its main approach and considered an effective way to resolve the problems of NRCS. Though the survey, I can summarize the result as follows.

First of all, UHIS resolve the situation of separation that exists between urban and rural areas with regards to the medical security issue and established a unified health care system. As elsewhere, the medical insurance systems for urban residents and NRCS in Zhanjiang were separated. Under the dual health insurance system, urban and rural residents pay insurance fees depending on very different standards and levels of operation where there is a huge gap between urban and rural regions. UHIS get rid of the identity boundaries and the split of dual health insurance systems by achieving a new pattern by which urban and rural dwellers pay for the same standards and enjoy the same level of treatment in health insurance.

Secondly, UHIS achieved a consistency between urban and rural areas in funding criteria, the insured benefits, and the level of treatment. According to the "one system, two levels, consistent between urban and rural areas, voluntary choice" principle, the standards of insurance fee were set of two grades of 20 RMB and 50 RMB (from 2010, to 30 RMB and 60 RMB) annually each person. The residents who participate in UHIS's family unit at the suitable pay grade receive care based on their income situation and the level of demand for medical services.

Thirdly, in the new system, social health insurance and business health insurance have been combined into one system. Among the premiums, 15 % of what was used went towards purchasing the large Medicaid insurance from THE CPIC and the remaining 85 % was used to purchase social medical insurance from Zhanjiang Social Insurance Bureau.

In addition, the insurance payment limit was been increased after the introduction of UHIS's the limit which was 15,000 before and was increased to 50,000 (for the premium of 20 RMB) and 80,000 (for the premium of 20 RMB). The social security Fund pay the reimbursement within 15,000 RMB and the CPIC pays the part which is over 15,000 RMB. Beginning in 2010, the reimbursement limit was increased to 80,000 RMB and 100,000 RMB without changing the medical insurance fee. The social security Fund pays the reimbursement up to 20,000 RMB and the CPIC pays the part which is over 20,000 RMB. Because the payment amount was increased significantly, the burden of medical expenses on insured participators was greatly eased.

Through the interview, I have known the specific implementation of the UHIS is as follows: The CPIC and Social Security Funds Administration (SSFA) established an evaluation system of designated hospitals for medical care quality. Firstly, the CPIC and the SSFA signed agreements with every designated hospital to establish the medical information common-sharing mechanism that states that the three sectors should share the information concerning the patient's condition and the medical services provided by doctors. Secondly, The CPIC sent a medical expert team of 77 members, some of them were present in four three-star level hospitals were used to inspect the other designated hospitals under the supervision of the medical services of each medical institution.

Thirdly, UHIS evaluated the designated hospitals satisfactory in a survey of medical services. The failed hospitals were given a warning and the seriously underperforming hospitals would be banned from operation. As a result, the hospital cost per capita was 4,746 RMB in 2007 before the introduction of the new system and it fell by 25.3 % to 3,543 RMB after the Implementation of the new system in 2009. The data show that the UHIS effectively prevented the phenomenon of excessive medical treatment and improved the level of medical services greatly. In addition, the patient's medical burden was also reduced. Besides checking the content of medical services, the inspection team also examined the patients to prevent malpractice and other such bogus treatment.

Finally, the UHIS set up a one-stop service and simplified the reimbursement process. Before the implementation of the UHIS and of the NRCS, rural residents needed to deposit all medical expenses, claim reimbursement and then complete the process to get back the amount of payment. However, the medical certificate had to go through the village, town and county authorities and it took a long time. As some villages are far away from the government station and transportation is inconvenient, rural residents were bemused by the entire reimbursement process.

After the implementation of the UHIS, third-party medical information sharing networks allowed the participators to be discharge of personal liability for medical expenses and let the remaining amount be settled between the insurers and the hospitals concerned. Settlement method facilitates greatly improved the situation for rural residents and the number of hospital visitors also appeared to increase. For example, the outpatient visitors to Chikan District Hospital maintained 60 % annual growth in 2009 and in 2010. As for hospital choice, the participators can choose not only the nearest clinic but off-site hospitals in the designated 182 hospitals within the province. Furthermore, relying on the nationwide network of cooperative medical institutions, UHIS can provide participators who are suffering from incurable diseases a nationwide health green channel and off-site referral system.

Furthermore, this method can significantly reduce administration costs through market operations. I also interviewed the Secretary of the Zhanjiang Social Security Fund Authority. As manager and operator of Zhanjiang urban and rural health care system, he described the saving in administrative costs. Due to the co-operation with a commercial insurance company, the NRCS resolved the shortage of social security funds and the confusion in management problems. Before the implementation of the NRCS, as elsewhere, the NRCS was separated from urban social insurance systems and there existed the situation of there being two management models, two teams and two networks in Zhanjiang. The NRCS was in charge of the health department and the social security agency was in charge of employee health insurance and medical insurance for urban residents.

After the combination had occurred, the Social Security Bureau managed to unify the management of NRCS urban medical insurance. According to the ratio of 1:8,000 standards between administrative staff and the number of insured residents, at least nearly 750 administrative staff was needed. However, the total number of permanent staff was only 224 over the jurisdiction of five counties and 6 districts together managing the insurance of work injury, unemployment and pension affairs. It means that the government needed to add more than 500 new members of staff. However, because of cooperation with the CPIC, the government has not increased staffing but reduced more than 20 staff members. The administrative cost of more than 800 million was saved annually and this has greatly reduced the administrative expenses of the Social Security Fund.

As a result, this model has achieved encouraging results. The field survey shows that in 2008, hospitalization rate for Zhanjiang was 3.7 % and it rose to 6.4 % after the implement of the UHIS in 2009 and nearly 8 % in 2010. The medical treatment received by the designated hospitals in Zhanjiang more than doubled in 2008. In 2009, the average per capita reimbursement rate was 51.2 %, compared to 38.6 % under the NRCS in 2008, it increased by 12.6 % points. The increase in hospitalization and the reimbursement rate is reflected in the insured person's benefit which has been extended. By far the interview in April 2011, the number of the participators in rural Zhanjiang reached about 4.88 million and the insurance rate was up to 98.4 %, the target of full coverage has basically been achieved.

6.4 Concluding Remarks

The twenty-first century is the era of ageing populations and ageing is universally global phenomenon. All the developed countries are facing the ageing problem and many developing countries are entering or will soon enter the "ageing society". China entered the ageing society in 2000. Because China is the world's most populated country and has the most ageing population, researching China's ageing problem is extremely important not only for China but also for the entire world. Meanwhile, because of inter-regional labor migration and other reasons, the degree of ageing is most serious in rural China, at the same time, the ageing process is growing fastest there. Because a social security system has not been established there yet, rural areas are experiencing greater pressure on them comparing to the cities. Solving the problem of rural security is an important issue for sustainable development in rural areas. As the medical and pension security are the two most important issues in rural areas, in this chapter, I have examined first the rural pension security scheme and secondly the medical problems.

In my analysis of old age pension security; I selected comparative research methods between China and Japan. On the one hand, Japan entered the ageing society at a mature period of rapid economic development. On the other hand, Japan also shows the small family-run model to be the main model in the area of agricultural management. This is very similar to China. Firstly, I compared the reasons for an ageing problem in terms of industrial structure change, labor migration and decrease in birth rates. Then I analyzed the pension systems in both countries and explored the sustainable path for an ageing rural China.

The government should establish a compound old age security system in rural areas. Firstly, the family should also play a fundamental role in the social security system. In the short term, because of the traditional culture, they have a major role to play an auxiliary to the one-child planning policy. The government should encourage and support the family with regards the problems of an ageing society. Secondly, a minimum living standard security system should be spread throughout rural China and both urban and rural residents should enjoy the same basic equal rights. Local governments should guarantee financial resources and the central government should shoulder the burden of partially or fully funding in under-developed areas. On that basis, the government should establish a unified pension system nationally and encourage rural residents of the economically developed areas to participate in commercial pension insurance schemes.

As for the choice of pension system, due to the high degree of in the population and becoming serious in rural China, sustainability is most important. As Japan adopted the "pay- as-you-go" pension system and is finding it difficult to sustain such a system in a time when the population is ageing, this system is not suitable for an ageing rural China. By contrast, as the peasants are self-employed, a fully funded or a partially funded model is a more feasible option. Besides, Countries should establish a rural pension insurance reserve fund, where part of the funds can be earmarked specifically to solve the rural ageing problems. The amount of reserve funds should be adjusted based on the degree of ageing in the rural population. As for nationwide distribution, it should be used flexibly based on the degree of need of each region and the fiscal ability of each local government.

Another extremely important issue for China's rural social security is the medical insurance system. Since the lack of health insurance, medical expenses become a heavy burden for Chinese peasants and their health is suffering as a consequence. To address the issue of rural medical security, the central government has proposed to establish a new-type of rural cooperative medical care system in 2002. Through the implementation of the new rural cooperative medical system, rural health care issues are initially addressed. But there are many problems. Firstly, the medical expenditure burden on participators is still heavy due to the moral crisis of some hospitals. Secondly, the administrative service is inefficient and based on high costs. Again, the new rural cooperative reimbursement system involves a lot of requirements and the reimbursement procedures are too complicated. In addition, the great gap in health care between rural and urban areas has not been eliminated.

Government failure is the root caused inefficiency in medical care system, so the central government has encouraged local governments to entrust qualified commercial insurance agencies to handle the new rural cooperative medical services and explore the establishment of supplementary medical insurance. The Zhanjiang government established an urban-rural unified health insurance system together with the Chinese People's Insurance Company in 2009. After 1 year of operation, it yielded good results.

To familiarize myself with their main practice, I carried out research in April 2011 and interviewed the General Manager Zhanjiang Branch. Firstly, the new system repaired the split between urban and rural areas by establishing a unified health care system. Secondly, through the combination of social and commercial insurance, the new system greatly increased insurance payments and eased the pressure on medical expenses for insured residents. Thirdly, they established a set of medical care evaluation systems and effectively monitored the designated hospitals, and so the medical service level was improved. Again, after the establishment of a one-stop service, the reimbursement procedure was simplified. Finally, the system significantly reduced administration costs through the market operation mode. Therefore, I believe that the pilot is an effective way to solve rural medical security problem.

References

Han F (2010) The rural old-age security function of land. Intellectual Property Press, Beijing, p 177

Leibenstein H (1975) The economic theory of fertility decline. Q J Econ 89(1). http://ideas.repec.org/a/tpr/qjecon/v89y1975i1p1-31.html. Accessed 26 Feb 2012

Li LQ (2009) The new rural cooperative medical system. People's Publishing House, Beijing, p 38

Liu JQ (2004) New theory of population economics. Southwest University of Finance Press, Chengdu, p 45

MHLW (2012) Report on pension fund management. http://www.mhlw.go.jp/topics/nenkin/zaisei/zaisei/data/data01/kokumin/kk-08.html. Accessed 7 Jan 2012

Ministry of Health of China (2004) Health statistical yearbook. http://www.moh.gov.cn/publicfiles//business/htmlfiles/zwgkzt/pwstj/index.htm. Accessed 25 Oct 2011

Ministry of Health, Labor and Welfare of Japan (MHLW) (2009) Vital statistics 2009. http://www.mhlw.go.jp/toukei/saikin/hw/jinkou/kakutei09/index.html. Accessed 18 Oct 2009

Nanjing Local Government (2003) Article 31 of "the population and family planning regulations in Nanjing". http://www.china.com.cn/chinese/PI-c/420721.htm. Accessed 14 Dec 2011

Qin BT (2009) A new view of Japan's childlessness. J Stud Mod Soc Cult 46. Graduate School of Modern Society and Culture, Niigata University, Niigata

Qin BT (2010) Analysis of labor migration in China on the Lewis two-sector model. New Vision in Economics Development and Environmental Conservation, No. 1, Graduate School of Modern Society and Culture, Niigata University, Niigata

Rong X (2011) The situation and problems of new rural cooperative medical insurance development, China Insurance News, 30 July 2004

The Chinese Central Governmental File (2011) No. 13, 2002, central committee and state council on further strengthening rural health work, Jiangsu health net. http://www.jswst.gov.cn/gb/jsswst/wjfg/sgfwj/ncws/userobject1ai12919.html. Accessed 9 Apr 2011

Wang YZh (2001) Analysis on the national role of health insurance in rural areas. Strateg Manag 3:14–15

Wu ChG, Zhou M (2004) Considering on the rebound of China's poverty population: where new poverty come from. Southern Weekend, 30 July 2004

Xiao LF (2007) Analysis of family pension in rural China. Observation in rural China. http://www.zgxcfx.com/Article_Show.asp?ArticleID=8252. Accessed 12 July 2011

Zhang WK (2002) Creating a new health situation in rural China. Health Manag Rural China 11:47–49

Chapter 7
Conclusion

In this book, I attempted to research the sustainable development in rural China. The three topics which have placed a main role in this research have been rural China's development in the areas of economics, ecology and social system. Based on the reality of rural China, there is too much factors related to sustainable development. This book has selected five factors of ecological agricultural production, industrial pollution, renewable energy, labor migration, and the provision of a welfare system which have been outlined in five chapters.

In Chap. 2, I looked at ecological agriculture through comparative research with Japan because the family-based agricultural operation model in Japan is closely related to that of China. Through the research into the Japanese government's strategies to develop ecological agriculture, the Chinese government should support the sustainable development in agriculture in the following respects: Firstly, as family-based land operations and small-scale agriculture are nationally applied features of rural China, and seeing that there is a lack of finances and technical know-how in the fields of ecological agriculture, the Chinese government should increase financial and technological support by promoting sustainable agricultural production. Secondly, as market demand is a prerequisite for production, the government should conduct some investigations and seek to expand the eco-agricultural market. With the market mechanism as it stands, the government should help agricultural workers by promoting sustainable agriculture production, the distribution of products and the system by which products are sold. In addition, the China government should also establish a multi-level certification system for sustainable agriculture. This will not only establish agricultural brands but also contribute to alleviating poverty in rural areas.

To eradicate poverty in rural China, industrialization has been seen as an effective means. However, industrial production causes environmental pollution and has become one of the main hampers to sustainable development. Therefore, I looked at the industrial pollution in the Chap. 3. I selected a chemical company and investigated the source of water pollution by conducting a field survey. To

B. Qin, *Sustainable Development in Rural China*,
DOI 10.1007/978-3-662-46476-2_7

resolve the environmental pollution problem, the government should change the way that economic is achieved by changing its "treatment after pollution" approach and establishing pollution prevention and a system of control. Regional economic development should be based on the appropriate regional environmental agency fully carrying out its duties to prevent pollution from urban to rural areas. As for the examination of site-settle industrial production, the establishment of highly-polluting projects in rural areas should be strictly banned. Special attention should be paid to the prevention of the types of pollution that take a long time to resolve, for instance, groundwater pollution.

Secondly, the improvement of legislation and law enforcement are the foundations of pollution prevention and control. When we examine the global situation, we can see that industrialized countries have also experienced this problem of how to solve environmental pollution. These countries have finally achieved a level of economic development which is needed for appropriate environmental protection because they have established a sound legal system. Therefore China should improve its environmental protection laws, especially with regards to economic globalization, legislation, pollution indicators and should gradually move closer to the position of developed countries.

Thirdly, China should establish strict enforcement and effective monitoring systems especially in rural areas. For instance, if the companies cannot resolve their pollution problems through technical means, their production activities should be strictly prohibited. If local government officials condone such production activities by colluding with the companies which are polluting the environment, then they should be severely punished.

Fourthly, as local officials have played such a decisive role in the decision-making process with regards to economic development and environmental protection, law enforcement significantly depends on them. The central government should change the evaluation content of local government performance and view environmental protection as its most important task. At the same time, the central government should establish a mechanism for environmental protection which targets responsibility and accountability thus encouraging local officials to engage in environmental protection work. In addition, because environmental officials are subject to the local government for their financial support, the central Ministry of Environmental Protection law enforcement cannot be carried out through to its final stages because local governments have proven to be uncooperative. To ensure environmental enforcement is independent, central government should establish a special salary fund for an environmental protection system by making the officials economically independent from local governments.

Finally, the Government should increase its investment into environmental protection in rural areas. It is important to study and formulate environmental pollution prevention and control planning according to the specific types of rural pollution that occur, while it should be that pollution prevention technology development in rural areas should continue to be developed so as to improve environmental protection capabilities.

Along with industrialization and the improvement in rural residents' living standards, electricity consumption has increased and this has put pressure on China's electricity supply capacity. Because the energy structure of China is heavily reliant on fossil fuel energy, such as coal and oil, this is not only the main cause of serious air pollution and acid rain, but it is also not sustainable due to the limited supply of such resources. Renewable energy, especially wind power, is one of the effective means by which China can develop a sustainable energy supply in the future. Therefore, the Chap. 4 went into the development of such renewable energy sources.

The progressive use of wind-power is the result of government support. However, China's wind-power industry currently faces many problems. Firstly, a survey into China's wind energy supply and resource has not been comprehensive enough. Abundant wind energy resources are extremely important when selecting a wind farm site. In China, the current distribution survey of wind resources is very vague, and it cannot meet the requirements of wind power companies to make a cost/benefit analysis. The Chinese government should refine the survey and provide more detailed wind distribution information to potential investors.

Secondly, the domestic technology level for manufacturing wind power equipment is not up to world standards. Wind turbine manufacturing technology needs to be nurtured so that wind power companies can further reduce costs. China still does not have a large-scale wind turbine manufacturing capacity, and small turbine units are prone to failure. The Government therefore needs to increase its investment in research and development to support various wind power enterprises.

Finally, financing channels for renewable energy are narrow. For wind power projects, the loan repayment period from domestic commercial banks is typically only 5–8 years. On the contrary, wind R&D needs long-term investment so that the development of advanced wind power technology is not restricted. To conclude, the Chinese government should establish a special financial institution to fund wind power research.

In rural China, compared to the limited arable land, there exists a huge surplus in rural manpower. Labor migration from the agricultural sector to the industrial sector has provided the necessary labor supply for industrialization. However, a labor force shortage has emerged in coastal cities recently. Lewis's two-sector theory has been recognized as a suitable explanation for China's actual situation and has been widely accepted by Chinese scholars. Many scholars believe that China's economy has reached the Lewis turning point. In Chap. 5, I analyzed this issue.

In the Lewis two-sector model, the most important factor promoting labor migration is economic. With the migration of rural workers into the city, economics factor has played a big role in deterring of the income gap which exists between urban and rural areas. In addition, a wealth of employment opportunities and a greater pension of facilities such as parks, roads, public transportation, etc have contributed to the reasons when workers have moved into the cities. Besides, convenient medical institutions and higher medical technology are other factors which rural labors have considering when migrating to the cities.

Even though these factors have been the main driving force behind migrant labors moving to the cities, many factors still block the flow. In China, the Hukou system is one the most important reasons for this. Although the rural labors work in cities, their "Hukou" has remained in the rural areas. In the city, they cannot enjoy the same social treatment as urban residents, such as medical insurance, and even their children cannot enjoy equal education opportunities. Although they can enjoy the convenience of their medical services, but have to pay more than city residents because they don't have health insurance. For their children to go to school, they have to pay expensive school sponsorship fees. Moreover, their rural "Hukou" reminds them of their status as farmers, so they cannot have a sense of belonging to the city. Driving factors and preventing factors have been competing with each other. Initially, rural incomes have been very low, so the driving factors have overcome the preventing ones. The attraction of higher wages and the wealth of employment opportunities have led to the occurrence of the rural–urban labor migration.

However, after the migrant workers have worked in the cities for a period, they have saved some hard earned money and started to be can't stand the inefficient situation in the cities for them especially due to the Hukou system. As a result, this has led some rural migrants to return back to the countryside eventually. Meanwhile, for those who are living in rural areas and have the intention of migrating to the city, the preventing factors will gradually overcome the driving factors for migration. The occurrence of a labor shortage in coastal areas is not a typical and countrywide problem but has just appeared in some localized areas. Even the migrant worker's average wage has increased and it has contributed to the increased demand for goods and services in rural areas but this has not been due to an increase in the labor supply from agricultural sector.

In China, the gap between the rural areas and urban areas has widened not narrowed. The development of township companies has increased rural income. As for the central government, on the one hand, they should look into reforming the HuKou system in regions with a view to expanding its phasing out in all regions. On the other hand, they should encourage middle and small scale companies as they can ease the pressure of surplus labor in the agricultural sector and industrialize the country's economy. Besides, it is necessary to establish a free, fair labor market and to give migrant workers the same opportunities in employment. For enterprises in coastal area, they have to face their labor shortage problem and carefully consider wage increases. For profit maximization and ensuring economic development, many coastal businesses have moved or are planning to move to labor-abundant inland area. Therefore the coastal economy has to transfer the emphasis from labor-intensive to technology-intensive operations. For China's overall economic development, it is essential to develop technology-intensive industries in coastal areas and labor-intensive industries in inland areas as there is an abundance of labor. This is how China's can develop its economy in a sustainable way.

The flexibility in the labor market causing greater migration from the rural areas to the cities has resulted in an aging population in those areas. Because most parts of the rural regions have not established a social ageing insurance system, and the new rural cooperative medical care system is still in its pilot phase, compared to cities, the ageing problem has brought greater pressure to bear on the rural areas. In Chap. 6, I analyzed the pensions system in rural China. As medical and pension security are the two most important issues in rural areas, in this chapter, I have analyzed the issues of the rural pensions and then looked at medical problems in the second part.

In my analysis of old age pension security; I selected research using comparative methods in evaluating the situation which exists in both China and Japan. On the one hand, Japan entered the ageing society in a relatively mature period of its economic development. On the other hand, Japan has also taken up the small family-run mode in agricultural management and this is very similar to China. Firstly, I have compared the reasons for an ageing problem with regards the change in the industrial structure change of the economy, the migration of labor and decrease in the birth rate. Then have analyzed the pension's systems in both countries and explore the possibility of sustainability in relation to the issue of ageing in rural China.

The government should establish a compound ageing security system in rural areas. Firstly, the family should also play a fundamental role in the social security system at least in the short term, as this forms an integral part of the Chinese society. As an auxiliary to the one-child policy, the government should encourage and support the family concerning the problems of ageing. Secondly, a minimum living standard in such a security system should be spread across rural China because both urban and rural residents should be able to enjoy equal basic living rights. It should be the case that local governments should guarantee the scheme's financial resources and central government should take the some of the financial burden partially or fully in respect to more under-developed areas. On this basis, the government should establish a unified pensions system nationally and encourage rural residents of the more economically developed areas to participate in a private pension scheme.

In choosing a pensions system, due to the high level of ageing within rural China, the question of sustainability is most important. As Japan adopted the pay- as-you-go pension system and is now facing a difficult period in its sustainability, this system is not suitable for rural China. In contrast, as the agricultural workers are self-employed, a fully funded or partially funded model is a more feasible option. Besides, Countries should establish rural pensions insurance reserve fund which is part of the funds earmarked for solving the problems of an ageing rural population. The amount of reserve funds should be adjusted based on the degree of ageing in the rural population. As for the distribution on a nationwide basis, it should be used flexibly given the degree of ageing in each region and the fiscal ability of each local government to provide funds.

On the contrary, since the lack of health insurance, medical expenses become a heavy burden for Chinese peasants and their health is suffering from serious harm. To address the rural medical security, the central government proposed to establish a rural new-type rural cooperative medical care system in 2002. Through the implementation of the new rural cooperative medical system, rural health care issues have been basically addressed. But there are still many problems with it. First of all, the medical expenditure on participators is still heavy due to a moral crisis in hospitals. Secondly, the administrative service is inefficient and run at a high cost. Then again, the new rural cooperative reimbursement system involves a lot of requirements for qualification for reimbursement procedures are much too complicated. Therefore, the great gap in the quality of health care between rural and urban communities has not been eliminated.

The root cause lies in the government's failure, to encourage local governments to entrust qualified commercial insurance agencies to handle the new rural cooperative medical services and explore the establishment of supplementary medical insurance. The Zhanjiang local government has already established an urban–rural unified health insurance system together with the assistance of the Chinese People's Insurance Company in 2009. After 1 year in operation, it yielded some good results. In order to acquaint myself with their main practices, I carried out research in April 2011 and interviewed the General Manager of the Zhanjiang Branch.

Firstly, the new system helped to repair the split that exists between urban and rural communities by establishing a unified health-care system. Secondly, through the combination of social and commercial insurance, the new system greatly increased the insurance payments and eased the pressure on the burden of medical expenses for insured residents. Thirdly, because there had been established a set of criteria for evaluation medical care and an effectively system for monitoring had been put in, the level of the medical care had significantly been improved. After the establishment of a one-stop service, the reimbursement procedure was then simplified. Finally, the system significantly reduced administration costs through the application of a market operation mode. Therefore, I believe that the pilot is an effective way to solve rural medical security problems.

In this book, there are three main points involving economy, ecology and the setting up of social systems which are the most important for sustainability of rural China. On the one hand, economic development is the driving force of agricultural modernization and industrialization. On the other hand, environmental pollution and ecological damage is one of the consequences for these developing models. Meanwhile, the labor migration from the agricultural sector to the industrial sector has caused an ageing problem and has put great pressure on the social security system in rural China.

To protect the environment, first of all, we should control industrial pollution with environmental legislation, law enforcement and corrections to the bureaucratic evaluation system. Secondly, it is necessary to improve modern agricultural production methods, and the choice to employ more ecologically friendly forms of agriculture is way forward in ensuring a brighter future. In addition, China should break its reliance on fossil-based energy structure and lessen its dependence on coal and oil, thus to develop and utilize renewable energy sources is the future path that china should take in ensuring a path of sustainable development.

However, the development of not only ecologically friendly agriculture, but also the development of renewable energy sources requires financial and technical support from the government. Therefore, in order to protect the environment and ecosystem, it is important for the government to lay a strong economic foundation. In other words, it is necessary to develop the economy, and to do so, a process of industrialization is considered one of the most important means of doing it. One of the most important factors for a continued process of industrialization is the huge reserve of surplus labor. Those workers who have transferred from agricultural sector to the industrial and so have provided the necessary labor supply. In this process, the reform of the social system (the household registration system) has played a positive role. However, the slow-down in the flow of labor and the labor shortage in coastal areas have shown that the progress of social system reform cannot keep up with the pace of economic development and so meet the needs of the labor force. Given on the analysis previously, it is possible to identify some paradigmatic scenarios for the development of rural China.

7.1 Scenario 1: The Default Scenario

This is the default scenario; it is based on the increased trend of population and labor in rural China in the near future, with no mitigation policies being implemented to resolve it. The government will continually divide the urban and rural area with a variety of restrictions and halt the migration of population due to the ensuring employment in cities. The increase of the surplus labor force causes rise in actual unemployment rates and cause the deterioration of social order not only in the countryside but also in cities. To maintain the social order, a large number of police and security forces are required. This would put a heavy fiscal burden on governmental expenditure and the government would fall into a financial crisis.

On the other hand, in order to maintain high agricultural production and economic income, agricultural workers will continue to heavily use fertilizers and pesticides and this makes agricultural poisoning more serious. The demand of food security will increase the import of foreign agricultural products and this will force the government into a policy dilemma between protection of domestic agricultural self-sufficiency and ensuring food safety.

At the same time, heavy polluting enterprises will continually transfer into the rural areas and industrial pollution will be more serious. The pollution incidents will occur widely in the countryside and eventually cause large-scale death and disability incidents. Due to the collusion between polluting enterprises and local governors, pollution condition will not be resolved. Local residents will stop the industrial production of polluting plants; as a result, conflicts will cause personal injuries. In addition, industrial production and electricity generation for life will increase energy consumption but China's energy structure will continue to heavily rely on the fossil fuels, especially on coal. In order to ease inflation due to the increase of energy price, the government will decrease the price of coal. The acid rain caused by burning coal land productivity will decrease arable land productivity and seriously damage human health and even the whole ecology.

Low-income cannot guarantee the necessary nutritional needs of rural residents; combined air and water pollution, these will cause the health deterioration of the rural residents. At the same time, due to a lack of a medical care system and a low level of disease treatment, patients in rural areas who should be hospitalized cannot receive hospitalization as they cannot afford medical expenses. Eventually highly infectious diseases would occur and widespread and a huge number of people would die.

7.2 Scenario 2: The Sustainable Scenario

This is the sustainable scenario. As the population of China declines and the government adopts the free migration policy, the agricultural population will decrease. Because the eco-agriculture has been spreading over the rural regions and acid pollution has been controlled, arable land productivity will rise. This will contribute to the remaining in grain yield or even increases it slightly. Even some arable land has been converted to grassland and forests to improve the ecological environment. On the other hand, because the government has promoted ecological friendly agriculture and supported the agricultural workers and organizations, the use of fertilizer and pesticide has been significantly reduced and thus the agricultural product safety would be guaranteed. At the same time, due to the high price of the ecological agricultural products, the income of agricultural workers will rise.

Due to the accumulation of capital from agricultural production, local companies will emerge continually. At the same time, the industrial investment from urban coastal areas will increase. As the local government has implemented the strict environmental protection policies, industrial pollution will reduce in contract of the development of industrialization in rural China. In addition, the industrialization in rural areas would absorb considerable amounts of the surplus in agricultural labor force and contribute to the increase of economic development and thus improve the living standards of local residents.

At the same time, the government will carry out the development strategy of renewable energy and provide technical, innovative, and financial support, and the prices of electricity will fall, thus the proportion of the renewable energy in the energy structure will increase significantly. Due to the gradual change of the coal dependency system and the technological development in the use of fossil energy, air pollution and acid rain caused by coal burning will be controlled. As a result, renewable energy can meet the consumption of the production and human living needs.

Industrialization and engaging of the eco-agriculture increases income levels and the living standards thus ensure the essential nutrients of rural residents. At the same time, due to adequate medical care system has been popularized throughout countryside, the health of the residents thus can be guaranteed.

7.3 Other Scenarios

Besides the previous worst and best scenarios, there exist some other scenarios as the government may focus on some of the topics and adopt some policies while ignoring other problems. Policies makers are always more concerned with short-term view of problems than with the long-term one, so more attention is always paid to economic growth. If the government just pays attention to few fields, even some problems could be resolved in a short-time. They may occur again in the long-run as the relative reasons have not been solved totally.

Even the history won't be end in a short time, but the time to resolve the problems left for us is limited. I can say that the earlier the government adopts reasonable policies, the easier it will be to achieve the sustainable situation in rural China. In order to fully achieve the end result described for the second scenario, efforts should begin immediately in a wide scale in rural areas. To achieve the sustainable situation in rural China is difficult not only politically but also technically. Therefore, International cooperation on technology should be implemented as many developed countries have considerable advantage in some fields such as renewable energy. Furthermore, it is extremely important for Chinese government to learn other countries' success system and it is one of the reasons why the comparative analysis method was adopted in this book.

Through the analysis carried out in this book, I was able to see that the three areas of economy, ecology and the establishment of a social system are interrelated. Development which is comprehensive coordinated and sustainable is the right developmental path for rural China to take. It can be achieved not only through inter-generational (vertical) but also through inter-regional (horizontal) sustainability. It is an extremely important issue and has a positive impact on global research into the sustainable path of development for rural China due to its large population size, vast land area and the impact that China's development has on the global economy. This research is certainly not conclusive in just a few studies; it is a task that should be carried out over a long period by many more researchers. This book is just one attempt at addressing save of the issues raised and only a beginning in my own study into such issues which I hope to pursue again in the foreseeable future.

GPSR Compliance
The European Union's (EU) General Product Safety Regulation (GPSR) is a set of rules that requires consumer products to be safe and our obligations to ensure this.

If you have any concerns about our products, you can contact us on

ProductSafety@springernature.com

In case Publisher is established outside the EU, the EU authorized representative is:

Springer Nature Customer Service Center GmbH
Europaplatz 3
69115 Heidelberg, Germany

www.ingramcontent.com/pod-product-compliance
Lightning Source LLC
Chambersburg PA
CBHW060742150926
51501CB00003B/71
* 9 7 8 3 6 6 2 4 6 4 7 5 5 *